Fanny Morris Smith

A Noble Art

Three Lectures on the Evolution and Construction of the Piano

Fanny Morris Smith

A Noble Art
Three Lectures on the Evolution and Construction of the Piano

ISBN/EAN: 9783744783743

Printed in Europe, USA, Canada, Australia, Japan

Cover: Foto ©berggeist007 / pixelio.de

More available books at **www.hansebooks.com**

A Noble Art

THREE LECTURES
ON THE EVOLUTION AND CONSTRUCTION
OF THE PIANO

BY

FANNY MORRIS SMITH

PUBLISHED BY

CHARLES F. TRETBAR

STEINWAY HALL

NEW YORK

TO THE PUPILS OF THE MISSES MASTERS' SCHOOL,
WHOSE INTEREST INVITED AND ENCOURAGED THESE
LECTURES, AND BEFORE WHOM THEY WERE FIRST
DELIVERED, THE AUTHOR DEDICATES THIS BOOK.

ii

PREFACE.

PIANO-PLAYING may be defined as the act of choosing and exciting musical vibrations in a piano. Because this requires manual dexterity, the human arm and hand have been carefully studied. We have learned what part anatomy, physiology, and mechanics contribute to those nice calculations which insure artistic execution.

But is familiarity with the manipulation of the power end of a lever, and ignorance of the resultant motions beyond the fulcrum, knowledge? And which of us knows the exact mechanical operation of any one of our simplest touches?

Few pianists are able to hear what no piano-maker can escape hearing. The mind that planned the action possesses secrets of touch for which artists vainly grope. To the piano-maker we must go if we would learn to hear and to strike.

Most piano teachers pass the majority of their waking hours before a piano. Those who, like the author, have seen the number of the hours thus spent increase from thousands to tens of thousands, must have asked themselves, What is a piano that it should exact and exhaust the energy, the hope, the possibility of human lives? What right have we to spend our lives thus? None, if the piano be only a harmonious accident of the nineteenth century. But if it be a necessary product of our civilization, if it be an adequate expression of the higher instincts of humanity, then our years have been rendered to the service of our race, and our life, no matter how straitened, toilsome, or harassed by poverty, is in itself worthy and dignified.

Such thoughts had long been habitual to the writer when it became her duty to plan a series of lectures for the school with which she is connected. They induced her to discuss "The Nature of the Artistic Instinct," "The Fundamental Laws of Musical Criticism," and "The Noble Art of Piano-making."

The writer at once began to collect and study every available authority upon piano-making; but, after many months of incessant work, the difficulties of the subject proved more insurmountable in exact proportion as its fascination increased.

Piano-makers being, for obvious reasons, unwilling to discuss the details of their craft, these lectures seemed hopeless. As a last resort the author presumed upon a very slight business acquaintance with Mr. Nahum Stetson, of the firm of Steinway & Sons, and asked permission to see the Steinway factory and to question the workmen. She met prompt acquiescence in her wishes; but when, six months later, she returned to ask for practical instruction in the principles of the art, and admission to the factory for study, her request produced unmistakable consternation, and was granted with great reluctance. Pianomaking is an art, and Mr. Stetson's sympathy with the writer's desire to demonstrate this to her pupils ultimately led him to interest his partner, Mr. Henry Ziegler (long associated with the late Theodore Steinway in his inventions), sufficiently to give the desired teaching.

To Mr. Ziegler's splendid acquirements in the theory and practice of his art, his enthusiasm for it, and his artistic spirit, this book owes whatever merit it may possess. If there be anything genial in its temper or lucid in its statement, it is caught from him.

The other partners of Steinway & Sons subsequently showed the kindliest interest in the writer's plans, but warned her of their Quixotism. "The public," each one declared, "even the student public, will care nothing for so technical a subject." Mr. Ziegler himself expressly restricted all information to the principles of piano-making, and absolutely declined to discuss the relative merits of any piano whatsoever, even his own. This chivalrous resolution was so strictly adhered to that the materials for the sketch of Theodore Steinway and his creative work were obtained with great difficulty. The writer collected them, a sentence at a time, from the inventor's workmen, friends, and family, whenever possible, during the last three years of her studies. In this matter she is much indebted to Mrs. C. F. Tretbar.

It should be explicitly understood that this book contains nothing which has been prompted or suggested
by the gentlemen without whose generous help not a
page of it could have been intelligently written. Nor
are they in any way responsible for the author's opinions and statements.

"The Artists of Piano-making" was written to ac-
quaint piano-students with the inner life and the artistic
spirit of the men who have created the piano.

The intelligent interest which these lectures excited in
their first hearers led to their repetition before other
schools, colleges, and adult audiences, and now prompts
their publication.

NEW YORK, September, 1892.

CONTENTS

A NOBLE ART.

I.

THE EVOLUTION OF THE PIANO.

Each day he wrought, and better than he planned,
Shape breeding shape beneath his restless hand;
The soul without still helps the soul within,
And its deft magic ends what we begin.— *George Eliot.*

THE science of piano-building has long seemed to me like an enchanted castle, a castle full of secret chambers to which no mortal has found the key. Over its door are chiseled mottos: Goethe's "*In every work of art all, even the smallest detail, depends on the conception*"; Whistler's "*In art it is criminal to go beyond the means used in its exercise.*" These and many more are deeply carven there by generations of artists and architects and builders.

Being entered, behold a hall and noble names blazoned on its walls; those of sages: Pythagoras, Aristotle, Archimedes; those of inventors: Jubal, Dædalus, and Watt; those of students: Newton, Faraday, Helmholtz, Thomson, Linnæus; of travelers, of statesmen, poets, artists; and, vanishing majestic in the obscurity, the symbols of antique civilizations, — Babylon, Egypt; of famous cities, — Damascus, Bagdad, and Susa. Beneath that mighty dome gather forgotten heroes and shrineless gods, — Bel, Isis, Ashtaroth; there linger Olympus and Valhalla.

People are not wanting who deny that such a castle exists. They aver that I saw only a monstrous factory upon whose grimy door was scrawled, "No admission except on business"; that the sages were a fat manufacturer, bedizened with diamonds the size of marrowfat peas; an anxious salesman, and a glib advertising agent; that the manufacturer beckoned me aside and said, "The foundation of the piano trade is the ignorance and helplessness of the public"; that the salesman whispered in my ear, "The actual cost of manufacture is not the moiety of the price I ask; the large and profitable fraction depends but on the airy fabric of a name"; that the advertising agent gave me a handbill, upon which I read, "The cheapest materials have the most money in them; the cheapest material for the manufacture of a first-class instrument is printer's ink."

None the less my castle exists. There is not a piano-maker who does not know of it and reverence it. To open to you the lordly halls and chambers thereof is the impelling cause of these lectures.

In seeking to establish the presumption that man possesses a spiritual nature, it is usual to point out such of his characteristics as are reflections of divine properties. In nothing does he show his divine lineage more emphatically than in the arts; and among the arts common consent has placed music foremost as a means to the expression of all feeling, and to the fullest expression alike of passion and of religious instinct.

If this be so, it follows that the musical instruments he has invented must possess a peculiar significance. Each ancient nation has in fact left in these instruments a gage of its progress toward civilization, and, what is more subtile still, of its temperament. The vina of India, the koto of Japan, the Celtic harp, the Italian violin, the piano of Christian civilization, each tells the story of the race that thought and fashioned it.

> For of the soul the body form doth take;
> For soul is form, and doth the body make.

When a musician makes his début before the public, he must have spent at least five hours a day for seven years, or seven hours a day for five years, in acquiring the requisite technical skill. This is over ten thousand hours. What must be a man's feeling for an instrument which for ten thousand hours has been part and parcel of himself; to which he has so united himself as to use it as his aptest means of expression; to whose keener and enlarged powers he owes it that he, who by himself is but a plain and stammering fellow, tongue-tied, diffident, very likely illiterate, if he but tuck his fiddle under his chin, can carry all souls with him? To the irreverent and disloyal, what friend more treacherous than the violin? more revengeful than the organ? more tricky, tantalizing, shallow, and contrary than the piano? How, into this rattle-box, shall one infuse a soul? and lo! before the patient and reverent spirit the malicious sprite retreats, and nobility and harmony reign supreme.

When this artist emerges from his ten thousand hours, he must inevitably have undergone a change, mental and physical. Action and reaction are equal: the instrument upon which he has expended the mechanical labor of a galley-slave has, in turn, effected its own alteration in him. As long as he lives he will never be exactly like the other members of the human family. His muscles have become elongated, able for great and sudden contraction, and of great endurance. But these qualities have been gained at the expense of lifting strength. They are not quite human, but feline. His skin has the true feline nervous irritability. All the centers of nervous power and motion, voluntary and involuntary, have undergone similar feline change. His ear, delicate as a hare's, has become abnormally sensitive to sounds and jars; his capacity for physical and mental suffering a thousandfold quickened and intensified. So much enforced abstinence from human companionship has made him unfit to meet the jostle of life on the usual easy terms of give and take. His feelings have the acuteness and sensitiveness of childhood; his art of interpretation is the growth and training of impulse; and he has the passions of a man. If we reflect a moment, it is apparent that this effect upon body and soul has occurred through the ear,— through the vibration of the nerves of the ear. The very quality of the tone thus habitually listened to has had a definite part in the total effect.

When man was made of the dust of the earth, his Creator breathed the breath of life into his nostrils, and man became a living soul. Ever since that day he has been striving to fashion this and that, and breathe the breath of life into the works of his own hands; in none of them has he more nearly succeeded than in his instruments of music; and the one that has absorbed the invention, discovery, and art of all the others is the piano.

The indispensable parts of a piano are four: ACTION, KEYS, STRINGS, and SOUNDBOARD. But the combination is modern; each of these separate factors has had a history of its own. The generator of a stringed instrument is evidently the string; given the string, two other factors are needful,—the means of exciting its vibrations and the means of reinforcing those vibrations sufficiently to affect the air to any considerable degree, and so to make them " carry." Among all musical instruments, antique, modern, civilized, and savage, wherever the string appears, some contrivance, no matter how rude, is associated with it, to receive and strengthen its tone. The scientific word for such an appliance is *resonator*. The resonator of a piano is called a soundboard. In the most rudimentary instruments it is only a hollow gourd or a strip of wood. Thus the simple gourd merimba developed into the vina.

A nation must have made some progress toward civilization before it essays to draw music from a string. The primitive instruments of the world were a rattle and a whistle. In the Department of Dordogne, France (formerly Perigord), have been found bone whistles made of the first digital phalanx of a ruminant animal, perforated with a hole, evidently bored with one of the little flint knives used in the stone age. We know no older or simpler instrument than this of prehistoric man. Later in the stone age was invented a pipe with three equidistant finger-holes. One of these instruments, made of stag's horn, was found in a bone-cave near Poitiers. A cane whistle with five equidistant finger-holes still survives among the shepherds of Barbary. The use of a resonant bowstring—such an instrument as is still to be heard among the Demaras, a tribe of Southwestern Africa—arose naturally from the hunter's habit of fondling his weapons. When not using their bows in war or chase, these savages tighten the string nearly in the middle by means of a leather thong, thereby obtaining two sounds, distinct but so weak as to be scarcely audible, except to the player. Another neighboring tribe knows how to reinforce the tone by attaching to the bow a hollow gourd, open at the top; while a third has advanced to a suitable wooden soundboard, strung with more than one string. But the soundboard has had an independent development through the drum. The Indians in New Mexico drum on two logs of resonant wood of differing pitch, to keep time for their dances, and get a good deal of melody out of them.

Daudet speaks of a curious combination of a three-holed fife and a drum, still to be seen in Provence. I extract the description from Virginia Champlin's translation: ·

"It is a *geloubet*, the simple rustic flute which goes toot-toot, while the drum goes boom-boom! What a drum, my friends! Tears came into my eyes when I saw it uncovered — an authentic tabour of the time of Louis XV., at once touching and comical in its enormity, grumbling like an old man at the least touch with the tip of the finger. It was made of fine walnut, ornamented with delicate carving, polished, thin, light, sonorous, and seeming to have become supple with the tempering of time. Serious as a pope, Buisson slung his tabour to his left arm, took the pipe between three fingers of his left hand (you have seen the pose and the instrument in some engraving of the eighteenth century, or at the bottom of one of old Moustier's plates), and, operating with his right hand the little ivory-tipped drumstick, he touched the big drum, which, with its continuous sound, like the burr-r-r of a harvest-fly, marks the rhythm and makes the bass under the shrill, sharp chirping of the *geloubet*. Toot! toot-toot! Boom-boom! Bright sunlight and warm perfumes filled my chamber. I felt myself transported to Provence, down beside the blue sea, to the shade of the poplars by the Rhone; roundelays and serenades sounded under the windows. They were singing glees, they were dancing olivettes, and I saw the farandole winding under the leafy plane-trees of the

village greens, through the white dust of the high-
ways, over the lavender beds of the scorched hillsides,
disappearing only to reappear, wilder and more abandoned; while the tabour-player followed with measured pace, quite sure that the dance would not leave
him behind."

The first savage who, quietly listening, found his
bowstring sound louder when attached to a block of
wood than when simply stretched by his bow, crossed
the chasm between barbarity and culture and took
humanity with him into the world of art—that world
whose *raison d'être*, as I have said elsewhere, is man's
need of expression. The nameless genius who in each
isolated nation crossed and carried his people over
this chasm which separates barbarism—no art—from
culture—loving art—did it in this simple step. Hence-
forth everything good was possible, for people would
now listen. Rightly then has this benefactor of his race
taken his place in many a mythology, as the hero,
the friend of man mediating between gods and mortals.

> . . . . Thou shinest in man's soul a god
> Who found and gave new passion and new joy,
> That naught but earth's destruction can destroy.

This greatest of all mortal musicians—greatest because
he created something out of nothing—must have
plucked his string with his fingers; but observation
and genius soon added the third factor in drumstick
and plectrum. Most vanished nations and modern savages knew, or know, the use of resonator, plectrum, and
drumstick, and have discovered a more or less rudimen-

tary scale. The culture of each people has grappled with the problems presented by our three factors, string, resonator, and exciting means, and illustrated them in

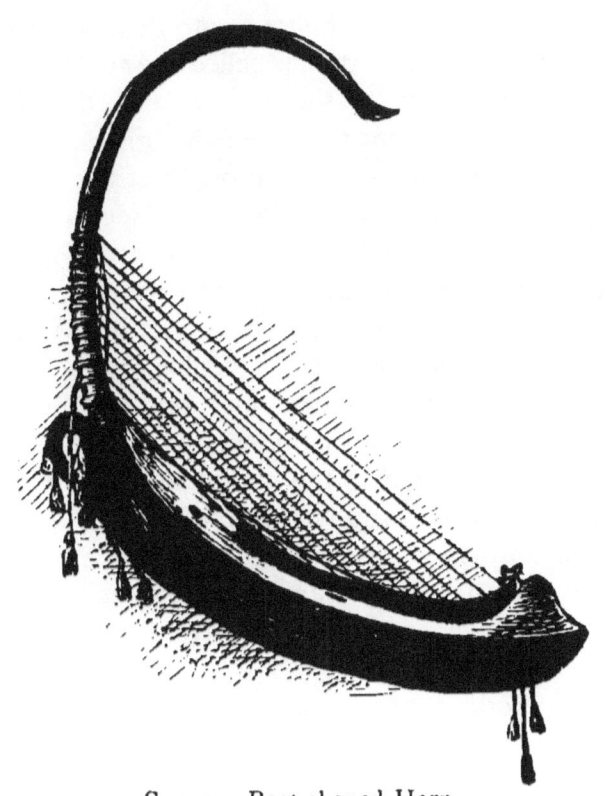

Soung — Boat-shaped Harp.
(From the Catalogue of the Crosby-Brown Collection. By permission.)

a hundred different combinations — each interesting, some most successful and artistic. Among the treasures of antiquity and of modern heathendom, we have mallets applied to rods of wood; strings reinforced by

2

hollow air-chambers, and twitched, struck, or scraped. A few principles have been apprehended and applied by humanity at large. For instance, that the resonator, or sounding-board, should communicate with its string at one point, at least, if the tone is to be at all strong. This point of communication is called the *bridge*; where the strings arise from the soundboard, as in the guitar, the point of attachment is in itself the bridge. In the violin family, where neither end of the string communicates directly with the soundboard, a special bridge is required. The soundboard has grown slowly from the stick and gourd of primitive man—reduced to a minimum in the lyre; doing double duty in the drum; single in the koto of Japan; double in the violin family; hollow air-chambered and bridged innumerably in the vina;—to its present dimensions in the piano. The historian of the piano can trace its forerunners in harp, viol, and lyre, before the altar of every great religion which the world has known. Persistent, like the boat-harp of Burmah, in by-places of the world still linger specimens of every stage of its development.

Among modern instruments one country, Italy, has loved and perfected the violin; France, the harp; Spain and Italy, as far as it has been done at all, the guitar; but the piano is the product and expression of the instinct of Christendom—no one race can claim its invention, no one country has given birth to its artists; and, as we shall see, no commerce less wide than that of to-day could bring together its materials. Nobles, artisans, poets, musicians, literati, have worked together to bring about its perfection, whose end is not yet. Cristofori in Italy, Marius in France, Schroeter in Germany, conceived the thought at the same time. Erard, a German; Backers, a Dutchman; Broadwood, a Scot; Southwell, an Irishman, were among its earliest inventors. Pape in France, Babcock and Chickering in America, are names that must come up as one considers its history. England, through Gray; Italy, through Cavallo; Germany, through Helmholtz, have

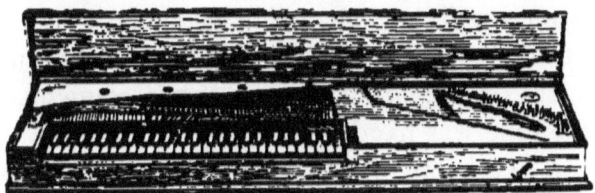

Clavichord.
(Steinert Collection.)

lent their science. Women like Nanette Streicher have worked in and owned its factories. Frederick the Great, at one extreme of aristocratic patronage, and John Jacob Astor, at the other, have equally interested themselves in it.

It seems probable that the immediate ancestor of the piano and violin is the monochord of the middle ages, once used to train the voice in convents; it was tuned with bridges pushed back and forth under the strings with the fingers. Originally its strings were stretched with weights hung at one end; later, tuning-pins were invented; and, finally (somewhere about the eleventh century), Europe came into possession of keys. From the East, thanks perhaps to the Crusaders and to the Moors of Spain, came the dulcimer, played with mallets, and the psaltery, whose strings were twitched with a plectrum, — which grew into clavichord and harpsichord. It was the dulcimer that, under the hands of Pantaleone Habenstreit, suggested to Schroeter the idea of making a compound of dulcimer and harpsichord — our modern piano.

Strings were first made of gut, but large instruments, like virginal and spinet, were strung with thin brass wire under small tension. The clavichord was constantly tuned in the act of playing by the pressure of the tangent on the string; but in the instruments with quills the tone was fixed, as in the piano.

At the beginning of the eighteenth century much thought had already been expended on the sound-board, which in the harpsichord was large and well

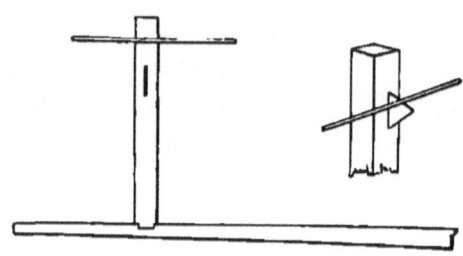

Clavichord action.

Front view of jack fixed in key (the tangent foreshortened to a line).
Side view of jack, the tangent in contact with string.

made, and on the various contrivances for exciting the strings, technically called the " action." This matter of action, the third factor of a stringed instru-

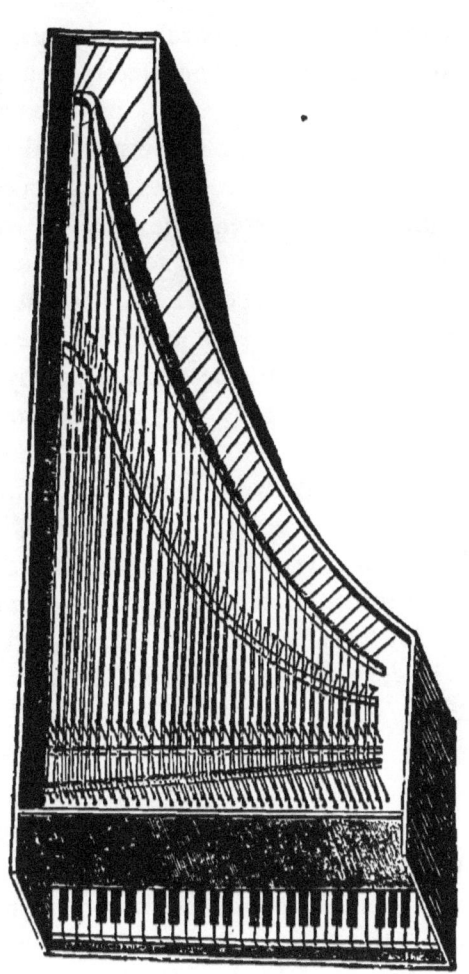

Harpsichord.

(From Rimbault's " The Pianoforte.")

ment, was the musical problem of the sixteenth and seventeenth centuries.

In Italy and Spain pure finger-instruments flourished. Associated with finger-action was the plectrum, a hook of steel that plucked the strings, and, by exciting a more angular vibration, produced a thinner and more penetrating tone. But the principles of successful "attack" were unknown; even the violin, whose manufacture was perfected in the seventeenth and eighteenth centuries, did not receive its present bow till toward the end of the eighteenth, although the bow, in an undeveloped state, had been created by the Aryan genius of India hundreds of years before the Christian era.

Not all keyed instruments were alike in action. In the year 1700, musicians were in possession of four keyed, stringed instruments, whose action varied according to its derivation from the plectrum or from the movable bridge. All four had certain features in common: clavichord, virginal, spinet, and harpsichord possessed the *soundboard*, which received and diffused the vibrations of the *strings*, which were fastened at one end by eyes hooked over pegs, and wound at the other over tuning-pegs fixed in a strip of wood glued to the soundboard, called the *wrest-plank*, from *wrest*, the key with which the minstrel tuned his harp; and the *keys*, not always exactly as we have them now, but divided and subdivided and twisted about to accommodate the exigencies of the old tonalities. These keys either carried fixed uprights armed with tangents,—triangular bits of brass or wood,—or else forced upward thin detached strips of wood, called *jacks*. There is an old *bon mot* of Lord Oxford, made while watching Queen Elizabeth play on the virginal, and alluding to Raleigh's favor and Essex's execution: "When jacks start up, heads go down." In the clavichord and monochord the key forced the tangent against the string, and thus excited its vibrations, and at the same time the tangent acted like a violinist's finger which shortens the vibrating-length of the string by pressing it upon the finger-board. As long as the tangent pressed the string, it would sound; but the lower end thereof being wound with tape to damp it, the whole was silenced when the tangent fell away. The tone of the clavichord, though thin and weak, was truly expressive. Bach loved it best. Dr. Burney says the son, C. P. Bach, absolutely contrived to produce a cry of sorrow and complaint from it. But

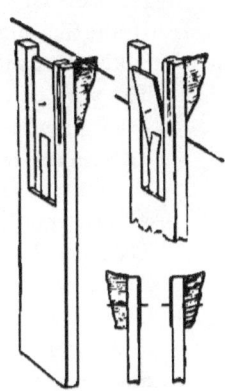

Action of a Harpsichord.

(In the possession of Mr. Bern. Bockelman.)

From behind. Jack in position below string. Jack escaped from string. Two views of jack-head from the side, showing damper and quill. The key is omitted, and also the jack-rail, which contains the collars in which the jacks work up and down at the tap of the key.

the stolid touch required such a cramped position of the hand as to preclude all execution, as we understand it. The fingers in attack were bent under at the ends, almost beyond the perpendicular, and released by drawing inward still farther toward the palm.

Shakspere has written the following sonnet to a lady playing upon the virginal—a curious instance of poetic license. He wilfully confuses the keys with the jacks.

> How oft when thou, my music, music play'st
> Upon that blessed wood, whose motion sounds
> With thy sweet fingers when thou gently sway'st
> The wiry concord that mine ear confounds,
> Do I envy those jacks that nimble leap
> To kiss the tender inward of thy hand;
> Whilst my poor lips, which should that harvest reap,
> At the wood's boldness by thee blushing stand!
> To be so tickled, they would change their state
> And situation with those dancing chips,
> O'er whom thy fingers walk with gentle gait,
> Making dead wood more blessed than living lips.
> Since saucy jacks so happy are in this,
> Give them thy fingers, me thy lips, to kiss.

Virginal and spinet were usually square, like tiny square pianos, or else five-cornered. The key, in their case, carried not a tangent, but a jack armed with a crow's quill, which, striking and then jerking past the string, gave it a twitch, set it in vibration, and then fell back from its own weight. No sustained tone—no dynamic modification—was possible, because, as the jack returned to position, the damper beside its quill silenced the string; but the harpsichord possessed considerable power. It was a sense of this lack of sustained and sympathetic tone that acted on man's longing for expression, and led to the invention of the piano. "Only through hammers could the harpsichord become expressive," said Schroeter.

As to names, clavichord was derived from *clavis*, a key; spinet — French, *l'épinette* — came from Spinetti, its inventor, but the word is influenced by *épine*, a bramble, from the points used to pluck the strings; virginal — Latin *virgo*, a maiden.

The action of spinet, virginal, and harpsichord was evolved from the plectrum. Their tone has been dubbed "a scratch with a sound at the end of it." One of them,

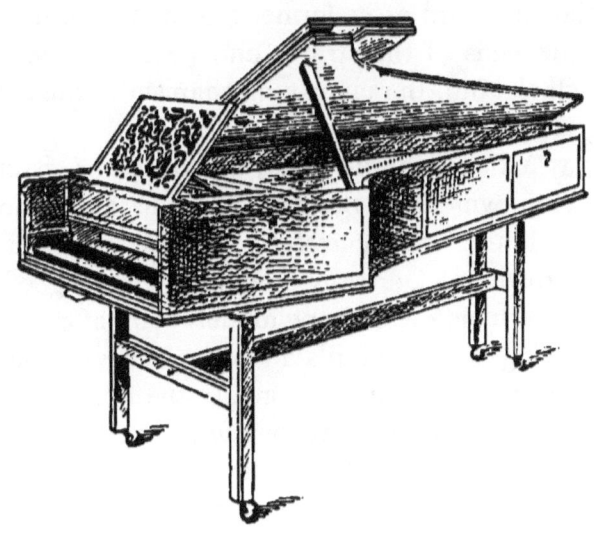

Kirkmann Harpsichord.

(Steinert Collection. By permission of Mr. Morris Steinert.)

— the harpsichord, — in the hands of Tschudi of London, was destined to still greater improvement, and long contested the field with the piano ; but all three are now of antiquarian interest only. The clavichord, however, is worth reviving. All these larger instruments were per-

3

fected between the thirteenth and eighteenth centuries, because music-wire was not made earlier, and (according to Mr. Hipkins) they preceded the piano only because the use of hammers involved a gap between the sound-board and the wrest-plank, that weakened the whole instrument and compelled it to wait three hundred years till a stronger frame could be planned. In the harpsichord the large soundboard is practically entire as to resisting power; but in the first pianos it dwindles to a third of its former size, and appears only under the ends of the strings on one side, exactly as in the clavichord, the clumsy mechanism of the action taking all the remaining space.

It was a harpsichord-maker, Bartolomeo Cristofori, in the employment of the Duke of Tuscany, who, in 1711, made the first successful piano; Marius of Paris and Schroeter of Germany—neither knowing of the other—producing less happy models soon after. Here is a drawing of Cristofori's action. It is very simple. The head of the hammer is small; the hopper is evidently a modification of the tongue in the harpsichord jack. But the idea of double levers is so correct that it has remained at the foundation of all delicate actions to the present day. This moving bit of wood is called the "hopper," because it hops forward and back. It is this motion which allows the hammer to fall away from the string so as to permit vibration. The escape of the hammer to its former position is technically called the "escapement."

Simple as is Cristofori's mechanism, it involved changes from the harpsichord model. The blow of the

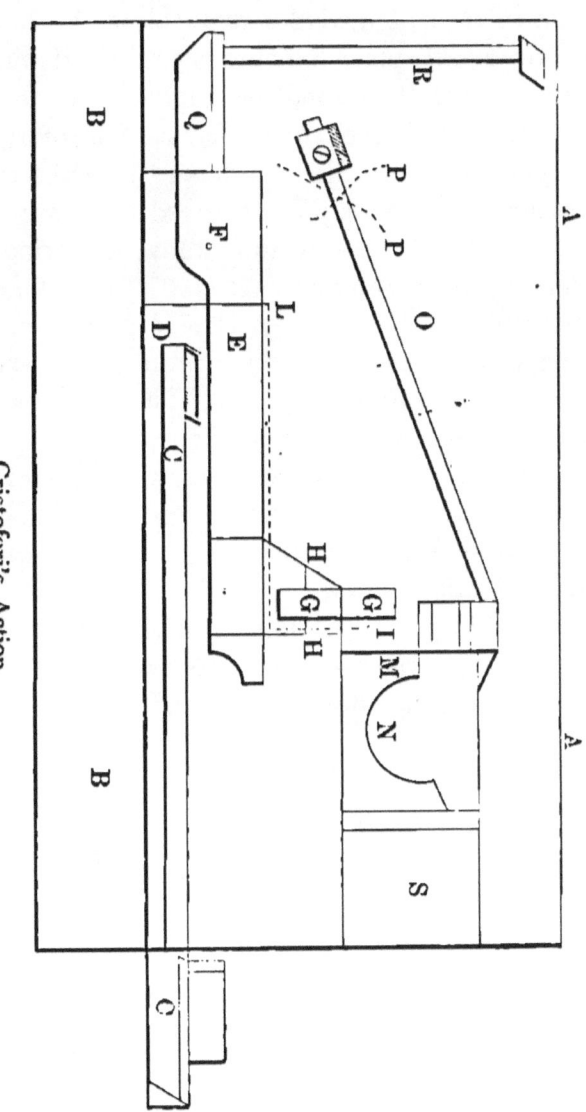

Cristofori's Action.

A, the string; B, key bottom; C, key; D, cushion on key; E, upper lever; F, center-pin of upper lever; Q, end of lever, bearing (under) damper on R, standard; P, P, crossed threads, forming hammer-rest; O, hammer; L, l, regulating springs of G, G, hopper, strung on wire H, H.; M, hammer-bar, in which turns hammer-butt N.

hammer necessitated heavier strings, whose increased tension in turn obliged the removal of the hitch-pins from their old place on the soundboard to a stiff rail of wood built around the angle side of the piano, and now called the *string-block*. Cristofori, as we have seen, cut down the soundboard to make room for the action. Ultimately he removed the dampers from the end of the key to a position above the strings,— the modern " over-damping." At least two of Cristofori's pianos still exist. Mr. A. J. Hipkins, who has played on one, pronounces " the action prompt and agreeable, and the tone not to be despised." The other two inventors were less successful, but " Schroeter, who perhaps never actually built a piano, foresaw the use of iron in the frame, and invented a ' resisting bar,' which, pressing firmly upon the strings transversely over the wrest-plank, formed a straight bridge."

It was Silberman of Dresden who made a popular success of the piano — thanks to Bach's praise, which awakened popular interest in it. He built the first in

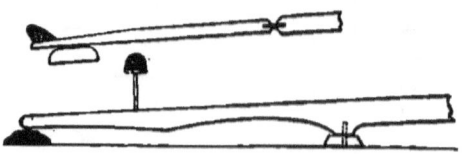

Zumpè's Action, showing form of jack known as " old man's head " (receiver of jack-stroke omitted).

1726, failed, was successful later, and sold one to Frederick the Great, about 1747, for the palace in Potsdam. Bechstein, Bülow's friend, a German piano-

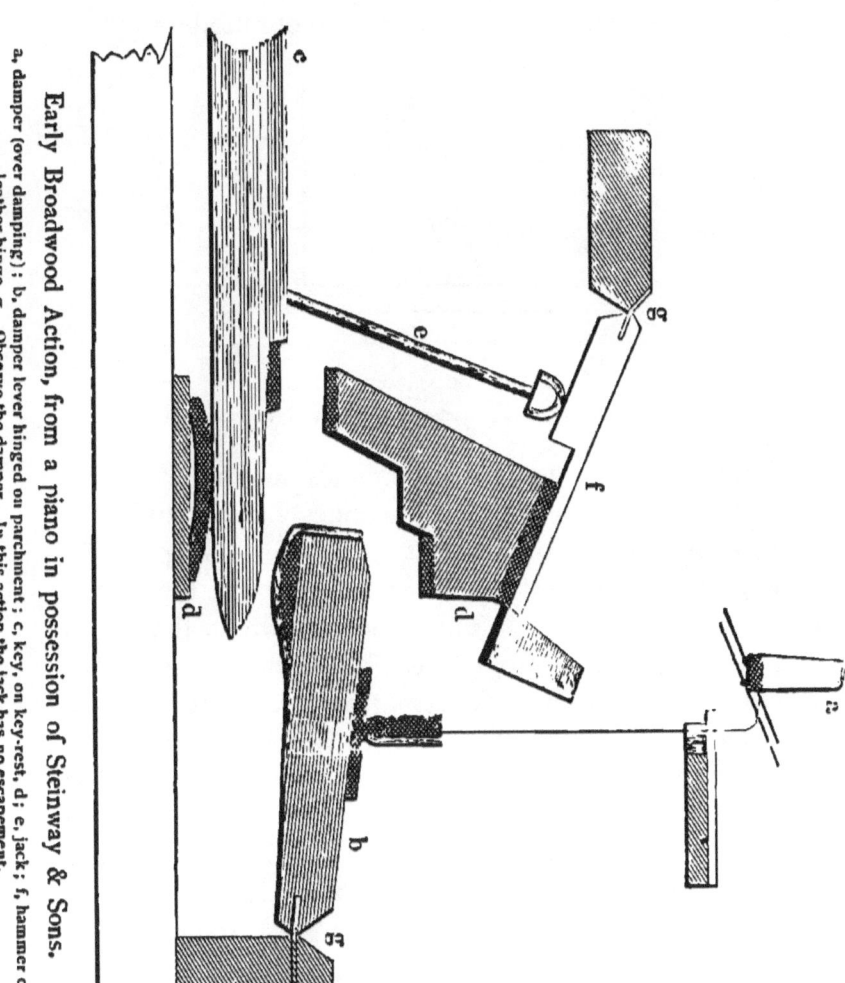

Early Broadwood Action, from a piano in possession of Steinway & Sons.

a, damper (over damping); b, damper lever hinged on parchment; c, key, on key-rest. d; e, jack; f, hammer on leather hinge, g. Observe the damper. In this action the jack has no escapement.

maker, has been allowed to examine it, and pronounces the action identical with Cristofori's. These pianos are shaped like the harpsichord and our modern grand.

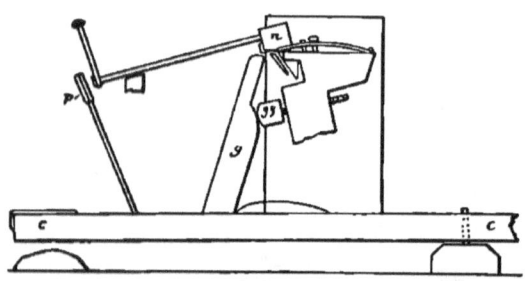

Broadwood's later Action.

c, c, key; *n,* hammer-butt; *p,* back-check; *g,* jack; *gr,* regulating button.

Within twenty years of Silberman's success, Frederici of Germany, followed by Johannes Zumpè, living in England, began to construct small pianos shaped like virginals. Possibly the latter followed the ideas suggested by Schroeter. The action shown in the cut is radically different from Cristofori's, and is sometimes credited to Mason, a clergyman and poet, the friend of Gray the Elegist. It is evidently derived from the clavichord. It would be a mistake to suppose that the invention of the piano passed from one man to another in one unbroken line. Once launched, every harpsichord-maker was busy with it, and worked independently. Mr. Morris Steinert possesses an instrument that is even a simpler adaptation of a clavichord. A bit of naked wood, immovably fixed on the key at right angles to it, strikes the string without hammer or escapement. Mrs. J. Crosby-Brown has in her collec-

tion a piano whose hammer-heads are simple bits of
wood at right angles to the shank. This piano possesses escapement and damping.

Americus Backers, a Dutchman employed by
Tschudi, the famous London harpsichord-maker, re-
turning to Cristofori's action, improved it by adding
a regulating screw and button before the jack to con-
trol the escapement. John Broadwood and Robert
Stodart worked with Backers. Their model here shown
consists of a key, a hopper, and a hammer. It is the
nucleus of the present English direct action, adopted
by Broadwood in London, Pleyel in Paris, and other
modern houses. This regulating button has passed
through a great variety of changes.

When Backers died he confided his grand action
to John Broadwood, but Broadwood was busy with
Zumpè's square. Broadwood's improvements were in
the soundboard and strings. He put the wrest-plank

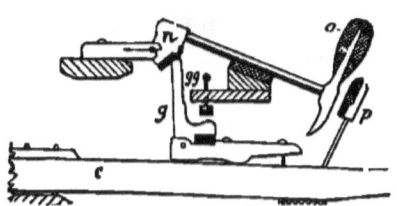

Another Regulating Button.
o, hammer ; *p*, back-check ; *c*, key ; *n*, hammer-butt ; *g*, jack ; *gg*, regulating button.

at the back, straightened the keys (hitherto crooked in
every direction), put in an under damper—*i.e.* a damper
on the same side of the string as the hammer—instead

of one falling from above upon the strings, and in 1783 invented the damper and sordine pedals still in use. He then turned his attention to the grand piano, and

Old Square Piano.
(In possession of Mr. Morris Steinert.)

called in Dr. Gray, of the British Museum, and Cavallo, who (*circa* 1788) calculated the tension of the strings (by a monochord), with a view to the equalization thereof. The result was that Broadwood divided the bridge on the soundboard, making a separate one for bass strings. It is a curious paradox to the laity that this great improvement of Broadwood is capped by Steinway's ring-bridge, almost a century later. The ring-bridge reunites what Broadwood severed.

John Geib is supposed to have been working for Longman and Boderip in 1786. Muzio Clementi was interested in this firm,—Clementi, the author of the "Gradus," and of our modern piano-playing. (It was Clementi who, in 1773, at the age of eighteen, published the first music ever written for the piano.) Geib returned to the hopper and under hammer of Cristofori. He regulated the motion of the hopper by piercing the blade with the screw invented by Backers.

The little piano in the music-hall where this lecture was first delivered, bearing the name of Geib & Walker, was made by Adam Geib, the son of this man, one of the twelve workmen who introduced

Broadwood Square.
(Steinert Collection.)

piano-making in England,—hence called the Twelve Apostles. Adam Geib and his brother John were in business in New York as early as 1802.

Mr. Spillane, in his "History of the American Piano-forte," has shown the close connection which has always existed between America and England in piano matters. The first piano known to have been built in America, according to him, was made in 1775 in Philadelphia by John Behrent. Ten years later one George Ulschoefer began manufacturing in New York. Boston came next in the persons of Bent in 1797, and Ben-

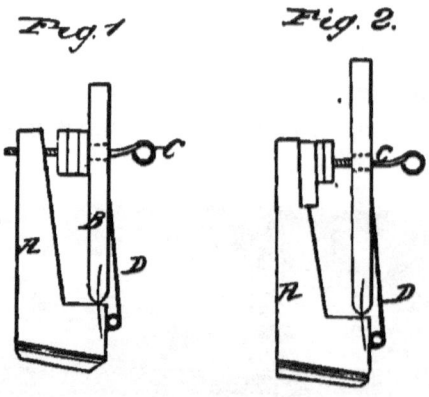

A Patent of Alpheus Babcock's, showing Regulating Screw in Fly.

A, jack ; *B*, fly in jack ; *D*, spring to fly ; *C*, regulating screw. Figure 2 shows Babcock's improvement in fly-cushion.

jamin Crehore in 1800; and lastly, Baltimore saw the Harpers well enough established to take James Stewart as apprentice in 1812.

Geib, the inventor of the grasshopper; Loud, the inventor of the overstrung scale; Stodart, who assisted Backers and Broadwood in perfecting the English grand piano; and Southwell, who made the first successful up-

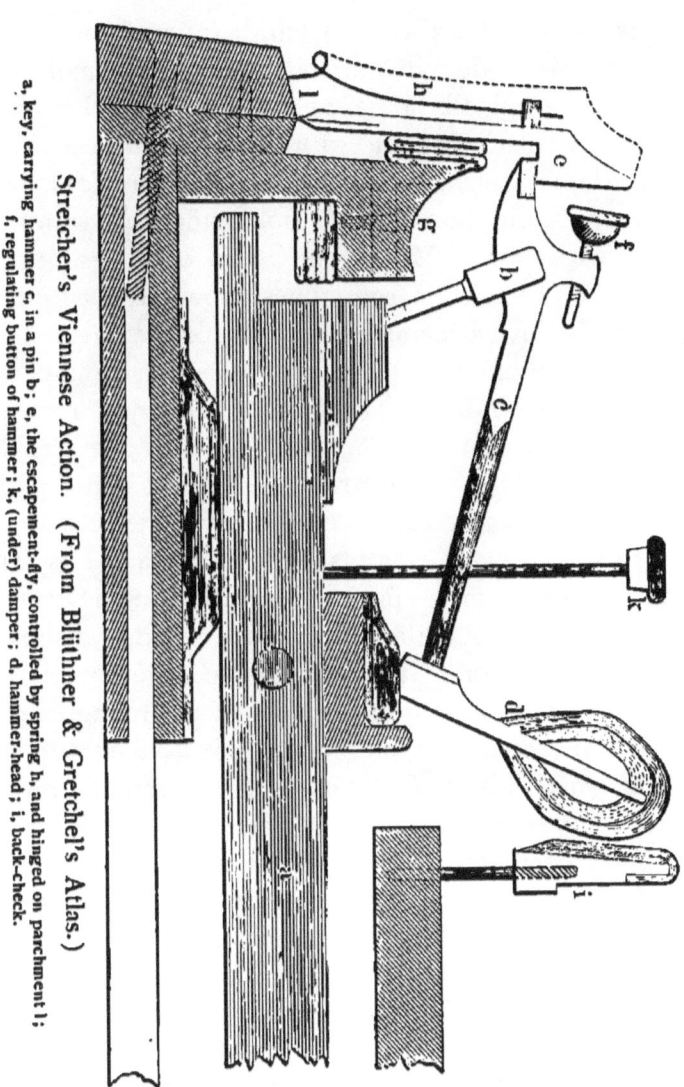

Streicher's Viennese Action. (From Blüthner & Gretchel's Atlas.)

a, key, carrying hammer c, in a pin b; e, the escapement-fly, controlled by spring h, and hinged on parchment l; f, regulating button of hammer; k, (under) damper; d, hammer-head; i, back-check.

right, all left descendants settled in America. Two of the firms founded by these families still exist under other names. On the other hand, James Stewart, who learned his trade with the Harpers in Baltimore, and who was originally a partner of Jonas Chickering of Boston, went to England with a piano of his own make as a sample; became the foreman of Collard & Collard, Clementi's old house; and took American traditions back with him. Meanwhile Stein of Augsburg invented a third action, sometimes known as the "pump-handle," which became the nucleus of the Viennese.

In 1789 Stein invented the verschiebung (una-corda) pedal. His son-in-law, Andreas Streicher, in 1794 improved Stein's action. The cut shows it as Streicher left it. In 1796 Sebastian Erard made his first grand piano, having previously constructed a square in 1777, two years after the first piano was made in America. His repetition mechanism, as patented in 1821, made its way slowly, but after Thalberg adopted, praised, and showed it, it gained favor. It is the favorite American action.

In the English action the butt of the hammer rests on a transverse bar; the hopper, pivoted on the key, works on this butt, first pushes the hammer upward, then escapes to its place. In the Viennese action the hammer itself stands in a pin on the key, and effects its escapement by pushing past the nose of a flexible bar behind. The Erard action involves at least three levers, including the key, the two constituting the action proper controlled by weights, springs, and buttons. Complicated as it seems, it is far more sensitive than either of the others, and can be better regulated.

The upright piano likewise passed through various stages. One of the three models of Marius was up-

Upright Hammer-Clavier.

(Steinert Collection. By permission.)

right. In 1787 John Landreth patented a grand turned on its side. Mr. Steinert has a six-octave André Stein (perpendicular grand) with four foot-pedals, dated 1779. But the first successful upright was invented by William Southwell, an Irishman from Dublin, who, after four-

teen years of experiment, patented his "cabinet" in 1807.
Four years later Robert Wornum came forward with
an upright strung diagonally, and perfected it in 1828,
after seventeen years' further study. In 1843 Wornum
patented another action, known as the "tape-check," so
that he must have spent half a century in experiment
in one department of piano-making alone.

To piano-lovers how much the dry, bare chronicle
of the invention of the piano means! How it speaks
of self-denying, patient investigation, of daring, of toil,
of hope, of tragedy. To understand this, one must
know the life of feverish enthusiasm and expenditure,
and remorseful despair, lived in torturing alternation
by the inventor. The idea that seems at the moment
of conception easy, practicable, and clear proves in
execution beset with difficulties, often insurmountable.
In the effort to conquer them, want, suspense, loss
often of entire fortune must be faced, and probably
felt. The story of Palissy the potter, whose means of
subsistence were consumed in search of the composition
of porcelain,—who tore up the floor of his cottage to
heat the furnace which fired his finally successful experiment,—is the typical history of invention.

"What do you know about piano-making?" said a
friend, who has himself worked mightily and successfully in another department of music. "I know! I am
a piano-maker's son. You cannot imagine how I love
and honor my father's memory. He was a true genius,
and his life was the pathetic story of continual disap-
pointment. He loved the art of piano-making. He
could take the tree out of the wood, and make every

part of the instrument with his own hands. But when it was done he was too sensitive to sell it. When people came to buy, if they seemed a little cold to its merits, 'Go away,' he would say; 'you do not understand my piano.'

"All day long the wind fluttered the papers hanging from the ceiling of our sitting-room — papers on which my father had drawn the action and construction of every piano he had seen; for, no matter how poor it might be, if it was by a new maker, he hurried to open, examine, and draw it. If good he learned its science, if bad he was wiser from its failure. Once up on the rafters, the designs never came down. No need. My father remembered them forever. But there they must hang, to the distress of my dear mother, who never ventured to touch them.

"No wire was even enough to suit his ear. He always made his own strings; and as with the wire, so with every other part of the piano. His whole life was a labor of love upon the ones he made; and, faithful as it was, and right as were his ideas about his art, he was never able to thrive in a little German town, where, when a man bought a piano, it was an event of importance to the whole village — something that happened not more than twice a year. When we finally moved to a city, strive as he would, he could never collect the necessary capital to begin a successful business. He was forced to eke out his means by tuning. I see him now, tramping from house to house, through winter storm and summer heat. When noon came he used to take a crust of bread from his pocket, and go to the nearest public house for a glass of beer to make his modest meal. He could not afford to buy the bread and the beer too. Often in need, we were never in debt. Honesty, industry, and self-denial were the rule of a life that in happier circumstances would have been famous. But his talent, and the love of his art that went with it, were sleepless until he died."

Such were the men whose history makes warp and woof of the noble art of piano-making. Love and hope, energy and struggle, and a broken heart! The cunning hand falters, and courage fails; the man dies; his name is forgotten. But the work he did, the impression of his self-sacrifice made on his generation, the tradition he helped to form, goes on forever. All his suffering is become a part of the immortal ideal he has followed. It is impossible to imagine an art created or perpetuated on any other terms. It is doubtful whether the very notion of an ideal can be separated from that of its companion martyr. And by our martyrs we can recognize and identify our art.

Very few of the more than five hundred patents actually taken out between 1779 and 1886, in England alone, ever brought the patentee more than loss and gray hairs; yet each experiment, each record, albeit of hope deferred and defeated, was a record of investigation necessary to our present knowledge.

For instance, a series of patents on glass, enamel, and mother-of-pearl demonstrated that ivory was the only entirely satisfactory substance for white keys, the reflection, the grain or lack of it, the degree of elasticity, being in each case a faulty departure from the ivory standard. A series of patents on piano-strings, single or corkscrew, of various metals or covered with various metals, formed the history of our present steel string, overspun with copper in the bass, to give weight.

A series of patents extending practically over more than fifty years, if we take in the harpsichord, of investigations lasting many times that period, upon mixtures of strings tuned in unison, and strings tuned to partial tones to make a good compound tone; plans for striking one string with two hammers, or several strings struck by one; plans for strings with extra bridges, and strings struck at various nodal points, are all links in the chain that has led to our present theory and practice: a practice in which no two great makers agree to-day, however much the experience of two centuries has added to its science and skill. The covering and make of the hammer, the position and form of the wrest-pins, the expedients for strengthening the framing, preparations for preserving and increasing the resonance of the wooden portions, have made and unmade many a fortune. Even such details as hinges and casters have passed from one inventor to another year after year.

The enormous cost of perfecting inventions and improvements in the piano has thrown such labor mainly upon the shoulders of the richest firms. But the manufactory is rare that has not one or two patents.

The piano has not made its way without conflict with prejudice; the French couplet quoted by Blondel tells the story:

5

Fier de ses sons moelleux que l'enfant sans peine
Avec un flegme anglais le piano se traine.

The *flegme anglais* is the rub, and explains why poets
like Voltaire should dub the clavier "a regular instrument for a brass-kettle maker." Philistines have been
busy with it from its birth. They have exhibited a passion for uniting it to such objects as "a harpsichord; a
set of Pan's pipes; a frame of musical glasses; an
Æolian attachment; a harmonium; an organ; a drum;
a harp effect; a music-box; a banjo; an angel lute;
wind and other musical instruments; a drum, tabor,
tambourine, and triangles; combination of piano, couch,
closet, and bureau with toilet articles; the music-stool is
constructed to contain a work-box, a looking glass, a
writing desk or table, and a set of drawers."

But the world's great piano-houses — Broadwood,
Chickering, Erard, Collard, Blüthner, Steinway—
have each worked with a genuine artistic ideal before
them, passing their talent and craft from one generation to another. Their purses and instruments have
always been at the service of music and musicians.
From the days when Nanette Streicher was the tender
friend of Beethoven, not only supplying him with his
piano, but caring for the vexed details of his living, to
the present, the history of modern music is largely the
history of their helpfulness.

It was Stein's piano that Mozart praised: "He
labors not for pecuniary interest, but for that of art,"
the young composer tells us. It was the piano the
Broadwoods gave Beethoven that Moscheles borrowed.

It was in Pleyel's concert-room that Chopin loved to
play, and in the direction of his genius has this house
felt its way.   Pleyel himself — we know his hymn,
"Children of the Heavenly King" — was a pianist and
composer; and so likewise was his partner Kalkbren-
ner, and Herz, his rival, who sought out many strange
inventions, and wrote countless variations.   Clementi,
who in teaching and inventing spent his life for the
piano, was a scholar and a gentleman.   His "Gradus" is
on our shelf; his "Sonatas" are dear to our memory. So,
too, Cramer, one of the original members of the house
of Chappell & Co., London, was one of the great piano-
teachers of his day.   To him we largely owe our legato,
and his studies are indispensable in the formation of a
pure and noble touch.   Moscheles writes down William Collard as one of the most intelligent men he ever
knew.   It was to Cramer's playing and Mr. Chappell's
hospitality that the London Philharmonic Society owes
its origin; and by Thomas Chappell, son of the latter,
were projected the famous Monday and Saturday popular concerts.   In America, the "Handel and Haydn"
Society of Boston owes it success largely to the exer-
tions of the Chickerings.   In New York, Steinway
Hall was for years the home of the orchestras of Dam-
rosch and Thomas, and of the Oratorio Society; and
all these organizations found in the house of Steinway
their useful friend.

The presence in America of the great pianists of
Europe, and through them the growth of musical taste
among us, have been made possible only by the exer-
tions of the great piano-houses.   That of Steinway

brought Rubinstein, Essipoff, Mehlig, Aus der Ohe, Paderewski, and a dozen others—names that recall to every American amateur a distinct epoch in his musical development. That of Chickering brought such artists as Bülow, Joseffy, and lately De Pachmann. Without the assistance of the great piano-makers of America no foreign pianist can be sure of making his expenses in a trip through our interior. If great artists sometimes play poor instruments, it is because without the backing of some piano-manufactory it would be folly for any manager to risk the uncertain patronage and fickle favor of an American public.

Piano-artisans are well paid, well fed, and intelligent. The greatest American makers care for the sanitary conditions and schools of their workmen. Nor does the help of the great piano-houses end here: struggling genius and forlorn and decrepit old age find assistance in them where their right hand knoweth not what their left hand doeth.

The history of the construction of the piano is woven of the lives of many men and diverse fortunes; but it is self-respecting, liberal, springing from human relationships of infinite sweetness, disinterestedness, and nobility. It has always displayed a restless striving after ideal excellence. Forgetting the things behind, its inventors have pressed forward to those which are before.

It has its courtly side in young Erard, whose boyish daring took him climbing to the top of the Augsburg Cathedral spire; who made his first piano in the house of the Duchess of Villeroi, and under her patronage; who, in his struggle with the fan-makers' guild in Paris, was sustained by the friendship of the nobility of France, and in his English contest for his patent, by the interest and aid of the nobility of England.   It has its homely, sweet romance in the tale of Broadwood, the workman in the factory of Tschudi, who married his master's daughter and became the ancestor of a line of inventors and improvers of the family instrument.

There is a higher touch of sentiment in the story of Streicher, who perfected the Viennese action.   Bred (according to Fétis) in an orphan-asylum, passionately fond of music, he finally came into possession of money enough to go to Hamburg and study composition under Emmanuel Bach.   But his love of Schiller's companionship led him to spend the little hoard in living in the poet's society.   Then he went to Munich, there to teach, publish, and fall in love with Nanette, the daughter of Stein, the piano-maker.   Working side by side with his wife in her factory, he improved the instrument, and when Nanette died grief cut short his own life.

Solomon, the Magnificent, gathered treasures from the ends of the earth to adorn his temple: fir and cedar from Tyre; "and the navy of Hiram brought in also great plenty of almug trees and precious stones; and the King made of the almug trees pillars for the house of the Lord and the King's house; harps also and psalteries for singers; and there came no more such almug trees nor were seen unto this day."   Temples like that of Diana at Ephesus, and

Apollo at Delphi, laid the whole world under contribution. Rimbault has gathered a list of woods and
valuables collected from every zone and continent to
compose the piano — that temple of the music of to-
day; its ivory, the wealth of the African village, the
cause of scheming, foray, and murder, brought pain-
fully through the black forests to the coasts,— a costly
treasure — every tusk the price of a human life. Its
ebony, the sign of an antique people and a tropic landscape; the riches of an extinct civilization, or a unique
race. How it calls up visions of Madagascar, holding
mimic court under her wonderful trees: of Ceylon, the
land of spice and pearl, still as in the day when Ezekiel
cried, "They brought thee, Tyre, horns of ivory and
ebony." See its mahogany-hunter, descendant of Mon-
tezuma, making his stealthy search in the pathless jungle
of Central America; hiding his cautious foot-tracks,
lest a broken twig or a handful of dead leaves betray
his prize; — or watch the hardy lumbermen of the Great
Lakes, felling spruce and maple in the cold northern
winter; opening a way to civilization; altering the cli-
mate of a continent; forcing us to a new lore of forestry
and a new science of political economy; — or study its
steel alloys, in whose initial experiments Faraday found
inviting field for his genius.

That sense of the mysterious, priceless riches of the
human soul — that new-found wealth of humanity
which has called our art into being — is of greater
worth than the fine goldsmiths' work of Ephesus, and
the oriental barter of Troy. If we can feel the throb
of human hearts in the vibrating pulse of its wonderful

body, our piano is a treasure indeed. Not under the lash, as the captives of Solomon's day wrought for Hiram, and the slaves of Ephesus toiled and sank beneath their load, were its materials collected. It truly bears the stain of every human passion; but it testifies to the tenderness of every social instinct. It is the offspring of courage, of daring, of skill, of industry. It is the nursling of the most ideal of fancies. Great painters have loved to limn its meadows with their grazing sheep. Travelers have threaded the wilderness to discover its precious woods, that grow from the equator to the arctic circle. How rare the birds that nested in those trees! Through what forests have paced its deer!

Its wages keep alight the hearth-fires of men of every tribe and nation. The patience of how many artisans' lives is here! The careful waiting and hope of how many women have followed the adventurous woodsman, the miner, the hunter, the sailor, whose lives are interwoven with it! How has the thought of those women and their children wrought in the workers skill and endurance, and brought them to a prosperous outcome! If it be true, as Hindu occultism seems to teach, that the human spirit is never altogether severed from what its soul has loved, or its mortal flesh has touched, then is every faith, every custom, every fear, every passion of humanity compacted in this body of music; and the playing hand of the artist needs not to create, but to recall.

II.

SCIENTIFIC CONSTRUCTION OF THE PIANO.

Geselle ist wer was kann ;
Meister ist wer was ersann ;
Lehrling ist Jedermann.

Theodore Steinway's Motto.

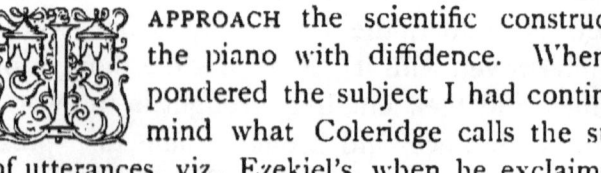 APPROACH the scientific construction of the piano with diffidence. When I first pondered the subject I had continually in mind what Coleridge calls the sublimest of utterances, viz., Ezekiel's, when he exclaims, "Can these dry bones live?"

To clothe them with living interest, it is most convenient to explain, piece by piece, the mechanism of an existing piano, made by one particular house.

An article of furniture may be described impersonally: for instance, a chair, as to its legs, seat, and back; a stove, as to its cylinder and shaker. Not so a picture or a statue.

I abstract from the "Critic" the following notice:
" For sale, a number of paintings, *all signed works,*
bearing the artist's name on a plate on the frame,
including one, 57 × 42, by Henry; one, 51 × 40, by
Bresciani, etc." Were the description more definite
you would read "Animals by Landseer"; "Landscape by Corot." It is the hint of the feeling and genius
of the artist that describes the work of art. So you must
say " Piano by Chickering or Steinway," and the name
separates a definite artistic ideal from the endless variety
of possible means and treatment,—separates an ideal
just as inherent in the genius of the inventor as does
the phrase " Interior by Gérôme," or " Evening by
Inness."

To treat of a general pattern of wooden plates and
iron girders would not initiate you into the charming secrets of the piano. We enter here the region
of perpetual choice and interdependent combination.
Here an ideal beckons, and a personal loyalty and
talent struggle toward it. In short, we deal with an
art. Now, an art presupposes an artist.

I have selected the Steinway piano. There are
others upon which a musician can express his feeling
with pleasure; their manufacturers have worked tire-
lessly, and spent their money freely and intelligently, in
perfecting them. But no house has associated itself in
closer fellowship with scientific research than that of
Steinway. Its members have been skilled handi-
craftsmen, whom the first scientific societies of Europe
have welcomed to their fellowship. The scientific
aspect of my theme can be most easily and naturally

6

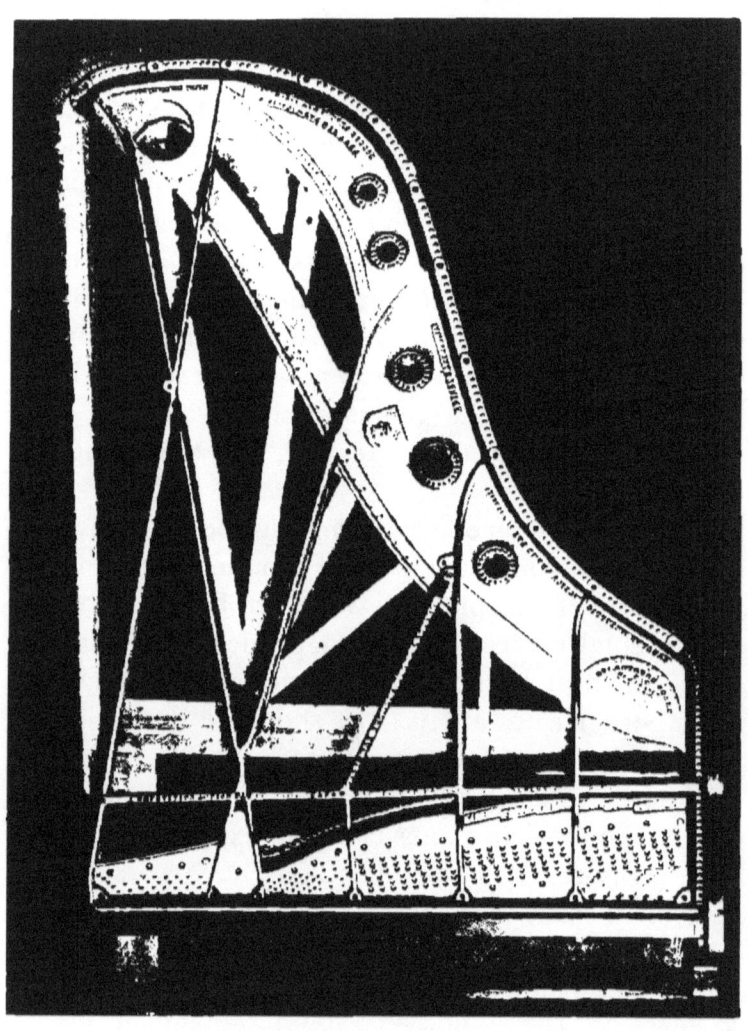

The Iron Frame, showing the Wooden Frame
(seen from above).

shown by reference to the piano originated by such minds. All the other great makers have confronted the same problems; all have arrived at more or less complete, though often widely different, solutions. Each house has worked along the lines of its own tradition, and, in seeking to preserve some excellence whose discovery was, perhaps, the initial point of its success, has more or less qualified all its succeeding investigations and improvements. A discussion of the variations of practice among the dozen great makers of to-day would occupy many months. If the piano were nothing but the box of strings it is crudely fancied, it would have no claim on the affection and respect of civilization, such as I hope to show to be its right. The loving toil which generations of men have devoted to its evolution is a touchstone to the strength of artistic emotion in the human family. Its progress, keeping step with that of science, makes it not only the child of this and that great manufacturing house, but the characteristic offspring of the learning of our age and civilization — of the genius of such men as Huyghens, Laplace, and Helmholtz — just as truly as it is of that of Erard, Bechstein, and Steinway. And I cannot insist too often that it is, above and beyond them, the fruit of Christianity, which, in teaching the infinite value of each human soul, gave each soul a right and necessity of expression such as it had never before possessed, and so taught humanity the art of Christian music.

Through the courtesy of Steinway & Sons I am able to show you these models of the essential parts of their piano: the rim and braces, resting on certain supports above the legs proper, viz., consoles and key-bottom; the soundboard, which lies upon this rim; the action, whose place, in a grand piano, is below and in front of the soundboard; and the iron frame, which not only carries the strings, but opposes the resistance of its tension-bars to their drawing-power. Beside these models I place the body of a Broadwood piano, made, as I judge from the five-and-a-half-octave keyboard, during the time of Haydn. (The Broadwoods claim the subsequent extension of compass [c to c] as theirs; though Southwell's six octaves [f to f] preceded them [Spillane].) If you will look at this Nunns & Clark, you will see an American piano probably prior to the patent of Jonas Chickering's overstrung square, and to the fan-scale of Henry Steinway. A glance shows the advance made since its construction. Its cramped strings, tiny iron plate, absurd hammers, hardly larger than those of the preceding century, all mark an earlier epoch of piano-history.

The sons of Henry Steinway who, in 1853, assisted their father to found his modest factory, had studied in different New York piano-factories, and represented what variety of traditional aim and method then obtained in America. Their piano expressed the best information of their day: was as much the physical embodiment of an artistic ideal as imperfect mechanical means permitted. The rich timbre of a violoncello sang in their souls—such a voice as we have heard from Parepa-Rosa. But their first actual instrument was but the germ of their piano of 1892. The wide gaps in the scientific and practical information of its builders

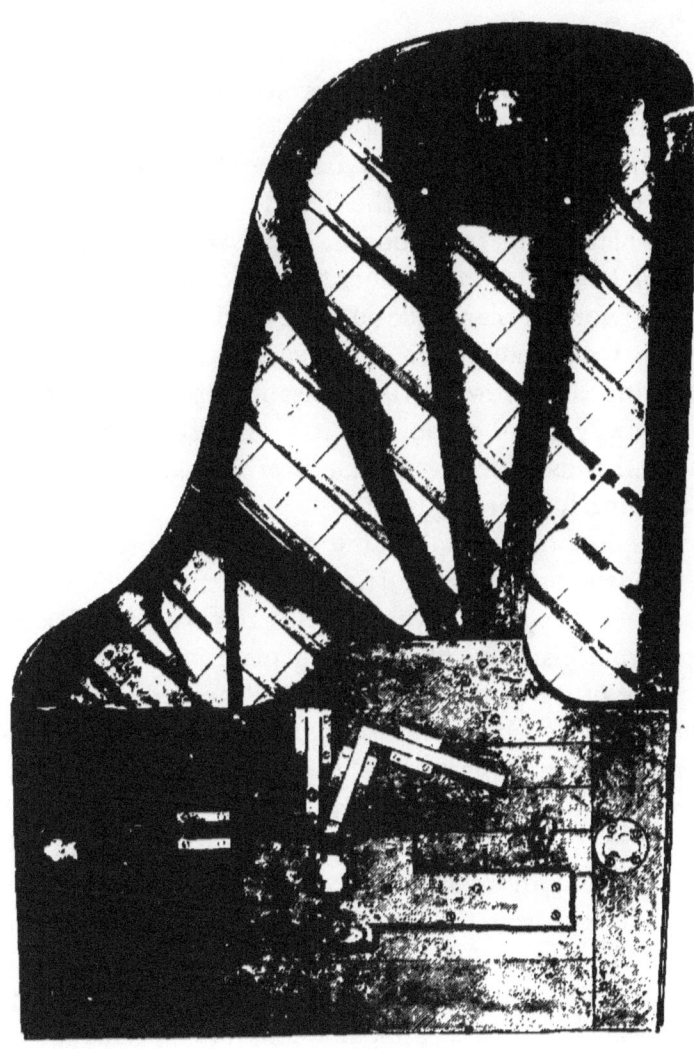

The Wooden Bracing, showing the Soundboard inserted in
its proper place (seen from beneath).

have been filled by incessant experiment and faithful work.

When a man sets about making a piano, he lays a sheet of paper the size of his projected instrument on a drafting-table, and draws a plan. On that plan he puts in its precise position every detail of the future instrument: the hammer-line, the lines for the keyboard, the wrest-plank, the soundboard, etc. When the plan is drawn, he makes and finishes the various parts to correspond; for the piano goes up like Solomon's Temple — hammer and chisel are not heard in it. There are, counting the screws, forty thousand separate pieces in a Steinway grand piano, and it is at least eight months in the hands of the workmen after the seasoned wood leaves the lumber-yards.

In spite of its metal frame and strings, the piano is a wooden instrument. It is easy to make the vibrations of the strings audible. But the artist builder selects from his mass of vibrating materials such as are capable of a fine tone-quality, gives these free play, and silences those which are dissonant. The natural vibration of the iron parts of a piano, very easily excited, gives rise to a high nasal timbre. I am indebted to Mr. J. Howard Foote for the information that in the manufacture of metal musical instruments it is very difficult to lower the absolute pitch of the vibrating material even as far as to the once-marked $\bar{c}$. The natural vibration-periods of metals are in the upper octaves. Even bells appear deep, not on account of their pitch, but of their loudness.

If you will observe the sounds of cymbals, tuning-

A Noble Art.

forks, triangles, or small bells, you will easily recognize in each this characteristic high metallic twang.

It is clear, then, that makers do not trust to the bronze and iron parts of a piano when considering tone. Metals are, primarily, factors of strength. Wood, on the contrary, while far more difficult to set in vibration, is capable of taking up and giving out the richest and most delicious resonance. Wood is not an uneven crystalline structure like cast metal, but an organic product of nature, possessing acoustic properties inherent in its cells and fibers. The art of piano-making lies in skilful choice and management of the wood which receives and diffuses the vibrations of the strings. Wood not only reproduces the sounds of other singing bodies, but is able to sing on its own account. I drop, one after another, these little wooden blocks of different lengths. You hear how the concussion excites in each a musical tone of definite pitch. The property of wood which enables it to conduct and impress tone upon other media is equally important. It is the vibration of its wood which gives the piano its flute-like quality. But before we can understand the whole value and use of the several parts constructed of wood, we must consider the action and strings, whose office it is to set it in vibration.

THE STRINGS.

THE part of a piano that is the exclusive property and invention of the maker is the "scale." This means the whole plan and proportion of the instrument, and also, in a narrower sense, its stringing,—which depends upon the behavior of a string under varying conditions. The contrivance I place before you is a monochord. The first ever seen in Europe was brought from Egypt by Pythagoras, some four hundred years before the Christian era. Upon this wire string I hope to show what is meant by nodes, partial tones, and transverse vibrations. With my violin-bow I set the string in vibration. You all hear its tone. The note, like all others with which we are familiar, is a compound formed by the union of several partial tones, all made by the vibrations of this single string. To quote Monsieur Marloye, "A string on which we act transversely can resound only on condition that it can yield at least two transverse sounds, of which the higher will depend on the part acted on, or on the mode of inducing the vibrations." Here are two paper riders. I place them upon the string, across which I draw my bow. Both are thrown off by the vibrations. I replace them, one at the middle of the string and one at a quarter of its length. I draw my bow across the string near the end, and both remain

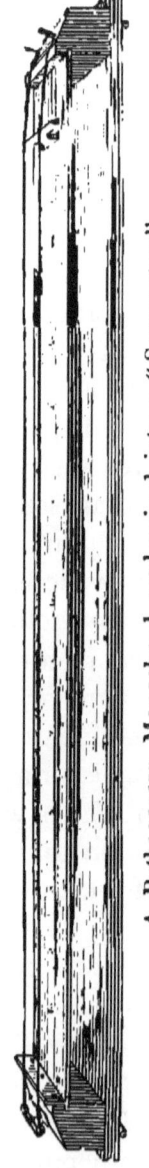

A Pythagorean Monochord, modernized into a "Sonometer."

in place. These quiet points of the string are called nodes. They are the points where it divides to form its partial tones. I damp it lightly on its center, and so withdraw the fundamental without eliminating the second partial tone, which you now hear an octave higher. The violinist calls such partials harmonics. By varying the place where the bow attacks the string, I can vary the formation of the nodes, and thus select the partial tones which they determine. A piano-string struck by a hammer not only vibrates as a whole, but separates itself into nodes and internodes, giving rise to a long series of partial tones.

I will play you a little study in partial notes excited by sympathetic vibration, quoted from Hans Schmitt's book on the pedal. I press middle *c* down without making any sound, and hold up its damper with the tone-sustaining pedal. Now I play the chords firmly, but staccato, and you can faintly hear an upper voice singing in harmonics upon the *c* which has not been struck at all. You must all hold your breath, for they are very delicate notes. The first six partials of a vibrating string comprise the notes of the com-

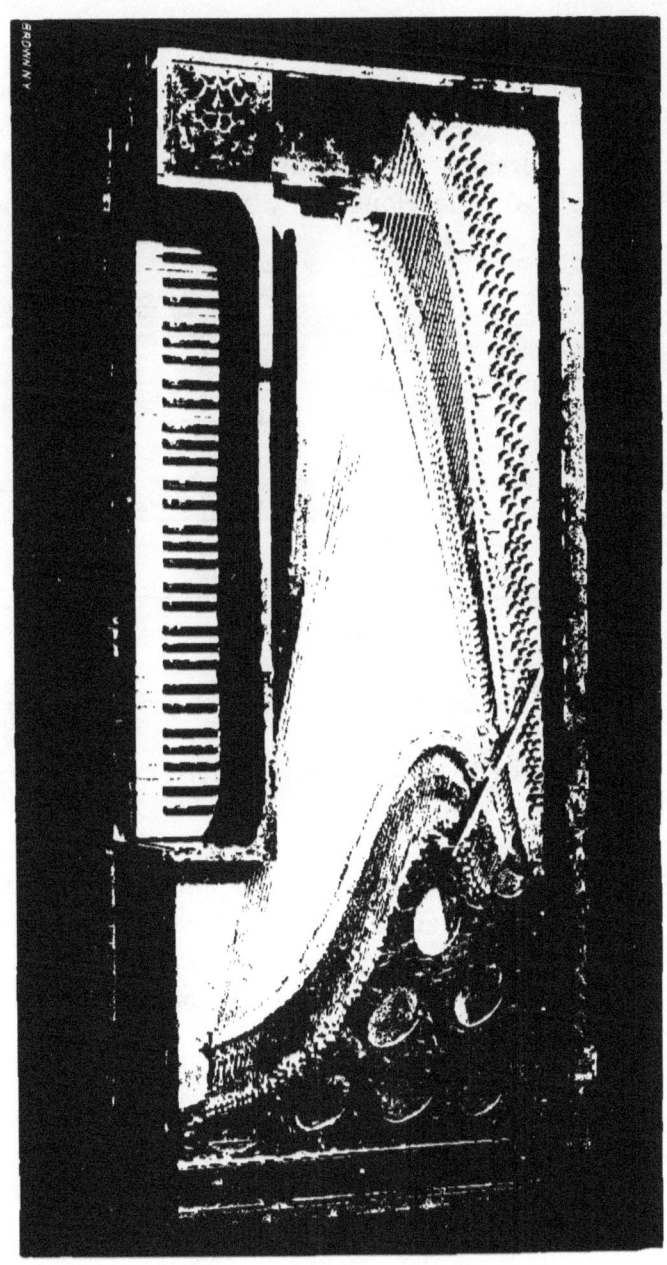

A Nunns & Clark Piano, in possession of Steinway & Sons.

Compare the soundboard, the soundboard-bridge, the hitch-pins, the wrest-pins, and the upper bracing with corresponding parts of the Broadwood and of the Steinway (shown elsewhere).

mon chord built upon its fundamental tone, and are therefore consonant. The seventh partial is somewhat dissonant; the ninth, tenth, etc., equally so. Helmholtz, therefore, thinks it desirable to have the first six partials well developed in a piano-string, but not the

seventh, and still less the ninth and higher overtones. Various things enter into the question: the thickness of the string, its length, its tension, its weight. The carrying property of a tone depends largely upon its constituent partials, and its delicacy still more; but before we arrive at the question of carrying, we must consider equality. There is a long series of measurements of the length and thickness of the strings between the bass and the upper treble. How shall they be graduated so that the tones make a perfectly even series from bottom to top? That difference in the tone-quality of the strings which lends itself to expression in a violin, would be fatal in a piano; but the life-struggle of the violinist to make his a and e strings agree in quality, is the struggle of the piano-maker throughout the entire scale. The tone of the loveliest violin in the world is hopelessly ruined if the e string be too thick for the a, or, as happens oftener, the a too thick for the e. When a violinist wishes to produce a robust tone, he uses rather thick strings and bears on with his bow; but if he wishes a pure, brilliant, and at the same time delicate tone, he chooses thin strings, and does not allow the weight of his arm to come upon the string the value of a hair. Such a bowing, depending on the strength and elasticity of the player's arm, will produce a tone which will seem free in the air. It is the study of the piano-maker to bring about similar conditions, and yet augment the power of his instrument to the utmost. The first consideration is, of course, to produce the requisite pitch without introducing several different qualities of tone. Observe the complications. The pitch of strings depends upon the rapidity of their vibration. This rapidity depends upon their length, thickness, tension, stiffness, and weight. Brass, silver, platinum, steel, gold, copper, are of different weights,—all have been tried. Since the weight of the string conditions the rapidity of its vibration, where short strings are required to produce a low tone, mixtures of heavy metals are made to reduce the pitch. Modern practice gives steel for the treble, and steel overspun with copper for the bass. Silver, which vibrates with a very clear tone, though useful for small instruments, has too little tenacity for pianos.

The pitch of a string varies inversely as its length. Imagine three strings: let one be twice as long as the other, but in thickness, weight of material, tension, exactly similar. The short one will vibrate twice as

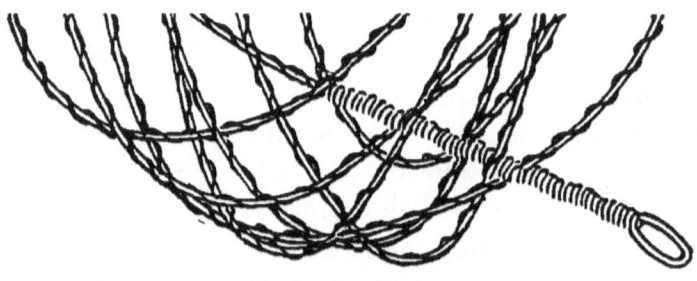

The Broadwood String.

The Steinway String.

The thickest bass string of the Broadwood Piano of 1809 compared with the lowest bass string of a Steinway Grand, 1892, natural size. Observe the old-fashioned corkscrew string.

fast as the long one, and thereby sound the octave of the latter. Again, let two be alike, except as to tension. If one be stretched with four times the stress of the other, it will vibrate twice as fast, and hence produce the octave of the looser string. Let it be twice as heavy, it will vibrate only one half as fast—that is, inversely as its weight. Now it is impossible to arrange the scale of a piano in such a way as to double the length of the strings for each octave. The pitch is therefore planned with reference to the mutual effect of length, tension, weight, and thickness,—a very nice problem indeed. If a string be too thin, it will not allow of a sufficiently heavy blow from the hammer, and the tone will be feeble. If the string be too much weighted, it will not yield its partials, and the tone will not carry or vibrate any length of time. Too heavy strings quickly lose their proper pitch. I have heard uprights so defective in this respect that their lower notes lost their definite pitch at a little distance from the instrument, and sounded like the noise of a drum.

The thickest bass string of the first pianos was thinner than the thinnest treble string of a modern instrument. It is the study of modern piano-makers to produce a scale with thick strings under high tension; because a string stretched to the limit of its tenacity yields the strongest transverse vibrations, and the purest and most brilliant tone. Now a string possesses different kinds of vibration, as Marloye long since discovered. It vibrates transversely when it forms the nodes and consequently the partial tones of which we have been speaking. But it may vibrate molecularly — that is, the sound-wave may move onward from molecule to molecule. This is the kind of vibration whose velocity Wertheim measured on telegraph wires — a vibration which, years ago, was utilized in the first telephone. The transverse vibration may be rotary. Marloye made it visible by attaching two colored riders to

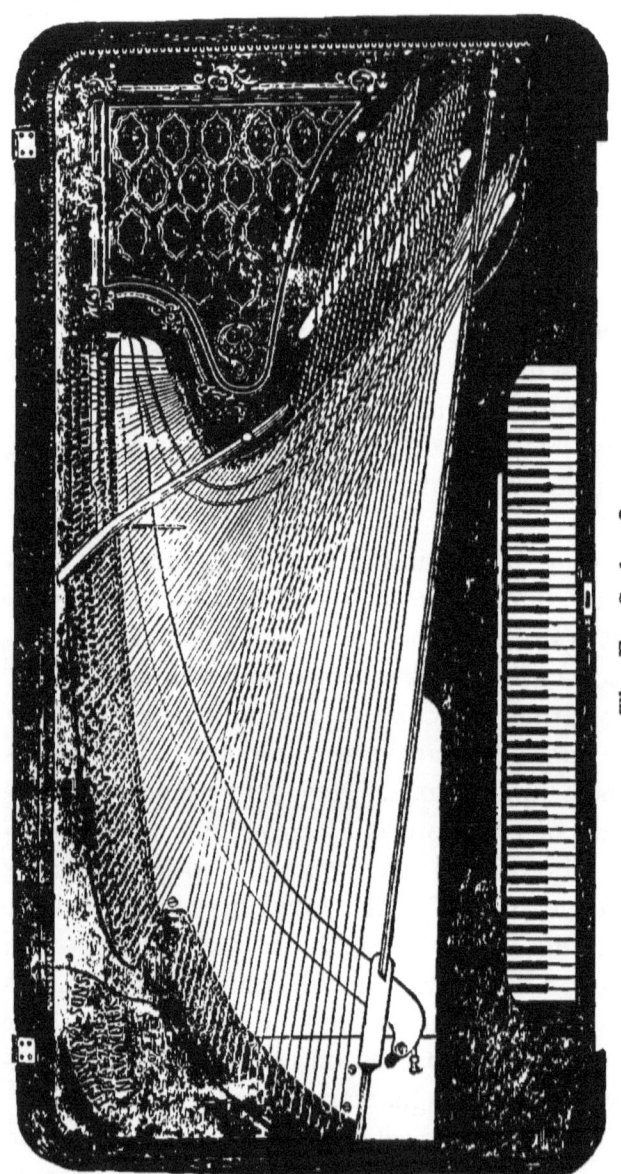

The Fan-Scale, 1855.

Compare the "full iron plate" and the "cross-string scale" with corresponding parts of the Broadwood and Nunns & Clark pianos. The Broadwood has no iron bracing whatever; the Nunns & Clark, a "half iron plate" at the right.

Permission of Steinway & Sons.

adjacent internodes of a vibrating string. He saw them revolve briskly in different directions. The rotary vibration of bass strings is sought by piano-makers. To promote it they arrange their hammers to strike obliquely.

Toward the close of the last century, John Broadwood calculated the tension of his piano-strings in order to equalize it, and so formulated the notion of a correct measurement of the string-scale. Thomas Loud of London made a cross-strung upright as early as 1802. In 1830 he was building cross-strung pianos in New York. Babcock of Boston introduced a cross-strung scale in this same year. Jonas Chickering patented an overstrung circular scale in 1845. The next step toward the present theory of stringing was taken by the Steinways, who divided their strings into webs, and arranged them in the form of a fan—the bridges in the center of the soundboard, the bass strings crossing the treble. This obtained two advantages: a better vibration for the strings, on account of their greater length; and a more complete vibration of the soundboard.

Nineteen years after the founding of their house, the Steinways patented their duplex scale, the complement of that fan-formed stringing which won their first triumph. We have seen the spontaneous division of the string into equal parts, each vibrating separately at the same time that the entire string vibrates as a whole—each vibrating segment of the string producing a partial tone.

Mr. Steinway arranged the transverse *capo d'astro* bar so that it took the place of agraffes in the treble,

8

and cut off a section of the string near the tuning-pin the exact length of one of the vibrating sections of the entire string. The bar touches the string, and forms a node; the end of the string thus cut off vibrates with the main part of the string, and, reacting on it, reinforces the partial tone to which its own vibration corresponds, and compels the formation of all the desirable intermediate partials. The other end of the string, between the bridge and the hitch-pin, is also an aliquot part of the main string. In this case the bridge forms not a node, but an impassable bar; nevertheless, this section of string vibrates sympathetically with the main string. We remember the researches by which Wheatstone showed that a substance in sounding vibration will set up similar vibrations in any other whose proportions will permit it to vibrate in the same periodic time. In this way the string behind the bridge reinforces the partial tone to which its length, thickness, and tension correspond. This scale, which rests upon the scientific investigations of Helmholtz, is one of the most beautiful applications of science to art that our civilization has seen.

Piano-makers agree that the point at which the hammer strikes the string greatly affects the tone; but they differ in practice. They hold that if the hammer attacks a nodal point, all the partial tones belonging to that node disappear, because the node, or point of rest, is thereby set in motion; but if the hammer attacks the center of an internode, it strengthens its vibrations. For example, if the center of the string is struck, the second partial, the octave of the funda-

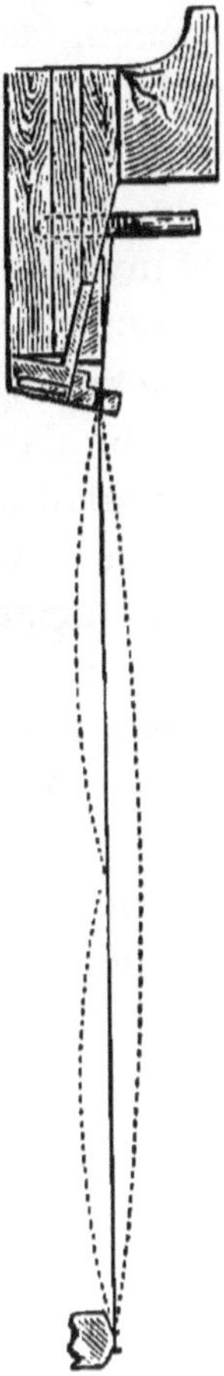

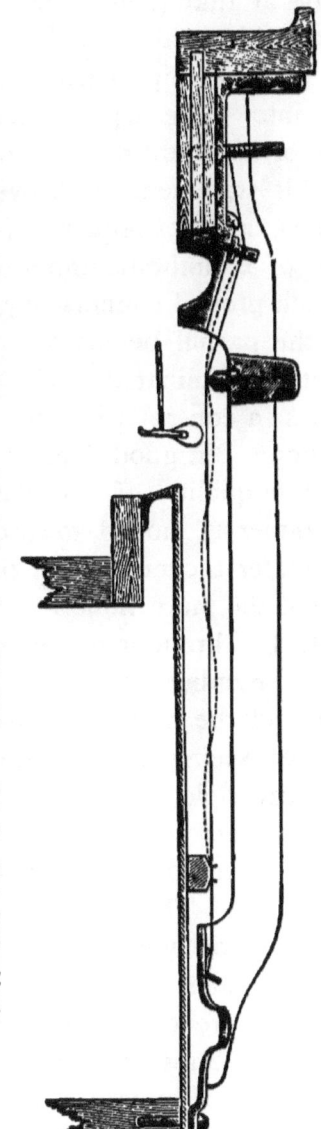

Fig. 1. Shows old wood construction with metal angle projection and agraffes, *without* duplex scale.

Fig. 2. Shows the original construction of the duplex scale, and the higher partial vibrations obtained by it.

From Steinway & Sons' Catalogue, by permission.

mental which has its node at that point, disappears, and also all other octaves of this partial, for that point is a node for them all; but the third partial, whose nodes divide the string into three equal parts, is strengthened, because one of its vibrating segments is excited by the blow. This statement works well in practice; but Mr. Hipkins, at the request of Mr. Alexander Ellis, translator of Helmholtz, undertook a series of investigations which proved conclusively that, though much weakened, the partial belonging to the node attacked was not totally eliminated. The point at which the damper comes in contact with the string is also decided by the nodes. But good damping depends even more upon the quality of the damper felt. The further the hammer is moved toward the center of the string, the greater the number of partial tones which are effaced, and the more hollow the timbre; but the nearer its attack is brought to the end of the string, the greater the number of partial tones which it permits to form, and the more piercing the tone becomes. I remember seeing a man show a poor violin on the same principle. He slipped his bow well toward the finger-board, and so cut off all the dissonant partials which would have formed if he had played near the bridge. The problem is to select (and provide a construction which will permit) that striking point which eliminates the dissonant partials and retains those which are consonant. Some American piano-makers allow almost no partials on their strings. The resulting tone is very much like that of a tuning-fork, tolerably loud, but not at all

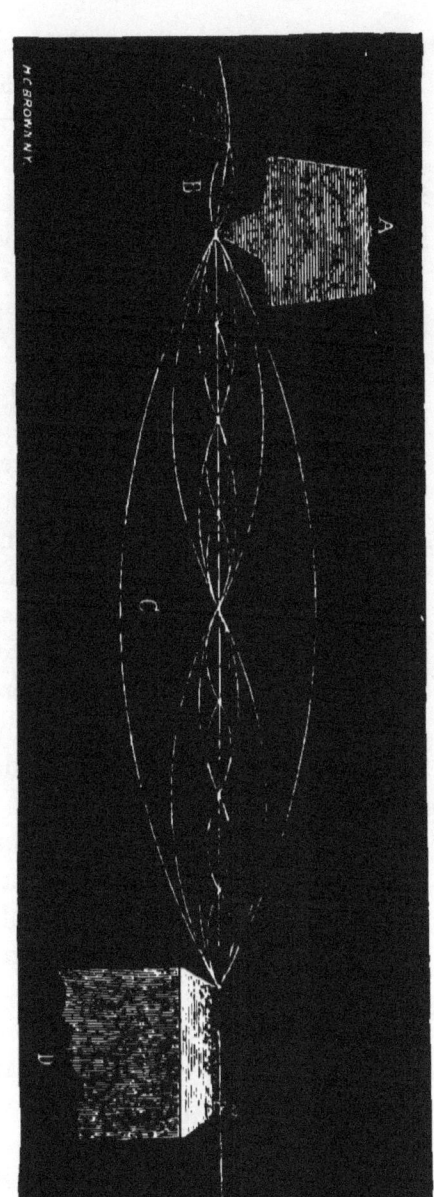

M C BROWN N.Y.

The Subdivisions of the String in the Duplex Scale.

A, Capo d'astro bar ; B, section of string cut off by it ; C, main string ; D, soundboard-bridge.

From Steinway & Sons' Catalogue, by permission.

sweet. Other makers place their striking-point near the end of the string; consequently their pianos, when a little worn, give a harsh, "tin-panny" tone, caused by the presence of a number of dissonant partials. These partials do not appear in a perfectly new piano, because the hammer is then comparatively soft. The usual striking-point is at one eighth or one ninth of the string-length — even less in the extreme treble. This matter is so important that there is an arrangement in Steinway pianos by which the entire mechanism of keys and action can be moved forward and back in the treble. The proper striking-point of the hammers on the string is thus secured with perfect exactness.

The size, weight, and covering of the hammer affect the tone in the most peremptory manner. An old maker once told me that the whole secret of piano-building was in the hammer. It is the subject of constant change and experiment in every progressive manufactory. If the hammer be too large or too soft, it damps off the string; if too light, the volume

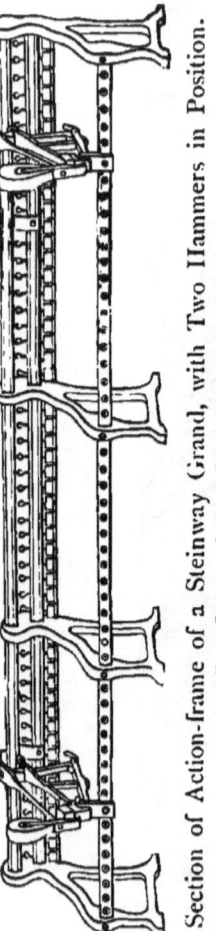

Section of Action-frame of a Steinway Grand, with Two Hammers in Position.

From Steinway & Sons' Catalogue, by permission.

of tone diminishes. If it be too hard, or ill regulated, so as to jerk the string, it produces an angular form of vibration, representing numerous dissonant partial tones, which manifest themselves more and more as the hammer grows still harder with use. The hammer must not only be well made, but must suit the scale and build of its piano. I once heard a Chickering grand which had been fitted with a set of large, soft Steinway hammers. It sounded as if it had a cold in its head.

We have seen, in the oblique position of the bass hammers, that the angle at which they attack the string alters the form of vibration,—therefore the quality of the tone. The length of time the hammer is in contact with the string, depending on the action, also affects the tone. According to Helmholtz, this should be just long enough to allow the consonant partials to form, and to eliminate those which are dissonant. Partials whose period of vibration is nearly twice the length of time during which the hammer is touching the string are specially favored; those whose periodic time is six, ten, or fourteen times as great are suppressed. The softness or hardness of the hammer conditions this result.

The art of hammer-making involves the elasticity of the head. Formerly several layers of skin were glued together by hand to make a hammer. America and France each claim as their discovery the substitution of felt for leather. At present the cutting and bending of the hammer-felt is done by machinery. The elasticity of the hammer is produced by the strain of the outer surface as it is brought into rounded form over the inner. Its hardness or softness varies according to

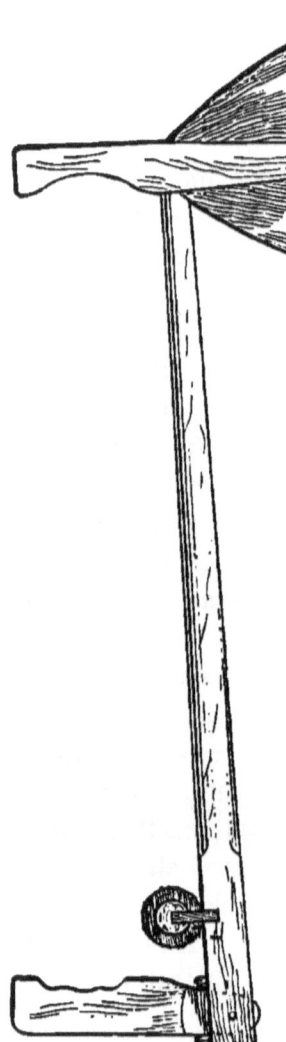

Hammer of a Steinway Grand, ¾ normal size.

the desire of the maker to favor the production of partial tones in his instruments, or to suppress them. No other part of the action wears out so soon; even comparative durability is therefore of the first consequence. Hammers are sometimes made of one layer of felt, sometimes of two, or even more: the outer softer than the inner both for durability and for mechanical reasons. It requires a constant and enormous pressure to bend the felt around the hammer-shank, with the rib of the felt inside. Among the great makers of piano-felt are Alfred Dolge of America and Weikert of Europe, who have spent their lives in perfecting the mixture of hair and wool which gives its durability and its velvet tone to a good hammer. It is the inner part which effects the greater elastic rebound; but the tremendous compression of the outer rim of the hammer gives great elasticity throughout. Neither the striking point of the hammer on the string, nor the duration of its contact with the string, nor the size and weight of the hammers among themselves, are uniform throughout the scale in any good piano. Different makers diverge widely in their individual practice. It is the art of the piano-maker so to vary the hammers as to compensate for, or qualify, the length, thickness, and tension of the strings which they excite. The longest

French Grand Hammer, ¾ normal size.

time of contact is a fraction of a second, and the variations are in minute fractions of a second. Upon the security of position and the accuracy of the motion of the hammers in their frame depends

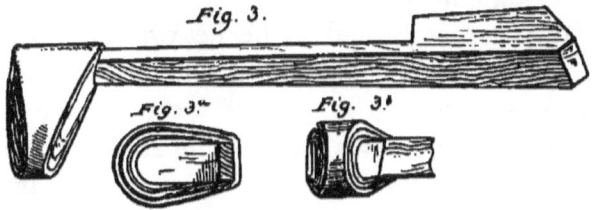

Fig. 3 : Broadwood Skin-hammer, 1809. Fig. 3*a* : End of Hammer. Fig. 3*b* : Striking Edge of Hammer, ¾ normal size.

the equality of the tone, no less than the durability of the action. Cheap pianos fail here most conspicuously. In the old system of securing the hammer-butts on

9

wooden frames, even the most careful preparation and choice of the wooden bars could not prevent their warping and swelling under climatic influence. The actions in Steinway pianos lie upon a metal frame, whose hollow tubes are filled with hard wood, forced in under great pressure. Into these tubes the screw-holes for the trains of actions are accurately bored. But the great expense of the expedient prevents its adoption in cheap pianos. Hammer-making, like the manufacture of piano-glue, is a business by itself; but first-class piano-makers make their own hammers. I have here a Steinway hammer, somewhat worn; a French hammer; and also the set belonging to my antique. The tiny, light, shapeless implement of the eighteenth century contrasts strongly with the large modern hammer from the Steinway grand piano. The hammer which is taken from the corresponding octave of a French action is smaller and lighter than the Steinway, and thus betrays at once the thinner string and slighter tension of the foreign instrument to which it belongs. The Steinway hammer is said to be heavy, and relatively so it is; but it is extremely light and elastic when its size is considered. The groove worn by the string acted, when in use, as a slight damper. If the hammer had been too thin and too hard, the same amount of wear would have cut near enough to the wood to produce an exceedingly harsh tone. You see the Scylla and Charybdis of piano-makers.

THE ACTION.

THE modern varieties of the *action*—that is, the mechanism which conveys the stroke of the pianist's finger to the hammer—were, in a more or less complete form, in the hands of manufacturers as early as the first quarter of our century. (Erard had patented his repetition action, the present favorite in America, as early as 1821.) All were to be greatly altered and improved by the application of the scientific principles belonging to leverage, striking energy, and friction. The touch is the most salient feature of the action. You see the key to be a lever; pressed by the finger, it raises the various carriers, hoppers, jacks, and pilots involved in the attack of the hammer. I have illustrated several primitive specimens of actions in a previous lecture: the English direct, whose hopper is pivoted on the key, or on a second lever raised by the key; the Viennese, whose key carries the hammer itself. The cut shows the repetition action of the Steinway grand, which is the Erard improved. The brilliance and purity of a pianist's tone depend very much upon the rapidity with which the finger attacks the key. To transfer this velocity to the stroke of the hammer without obliging a dogged pressure on the key, is the *sine qua non* of a perfect action; and this problem has been in the hands of piano-makers from the very beginning. But there are other complications. A thick string requires a heavy hammer and a strong blow. Makers have usually taken their choice between thin strings, small hammers, and light actions, and heavy strings and clumsy actions. The question also comes up in violin-making. A violin with a light elastic tone, that speaks quickly, requires little expenditure of strength in bowing, but it lacks nobility; whereas a violin with a robust and powerful tone demands a corresponding expenditure of muscular power.

It was the salvation of the Steinway piano that its makers sought tone before everything else. This being obtained, after unremitting experiment they perfected an action which satisfied all the desirable conditions both of tone and of touch.

The total weight, friction, and inertia of the action is overcome by the key-lever, by the weight of certain leads imbedded in one of the arms of the key-lever, and by upbearing springs.

The relative length of the arms of the levers; the relative positions of their fulcra; the evenness of the key-bed; the thickness and quality of the felt under the key; the strength of the springs in the action; the weight, size, shape, and materials of the various parts of the action; the angles of lift and stroke, offer nice problems in mechanics; so, too, the precise points at which the various checks come into play, and the manufacture of the checks themselves.

A perfect action transfers the blow of the finger to the hammer without loss of power or of velocity. Its touch should be delicate, yet elastic; regulated, as

Action of a Steinway Grand.

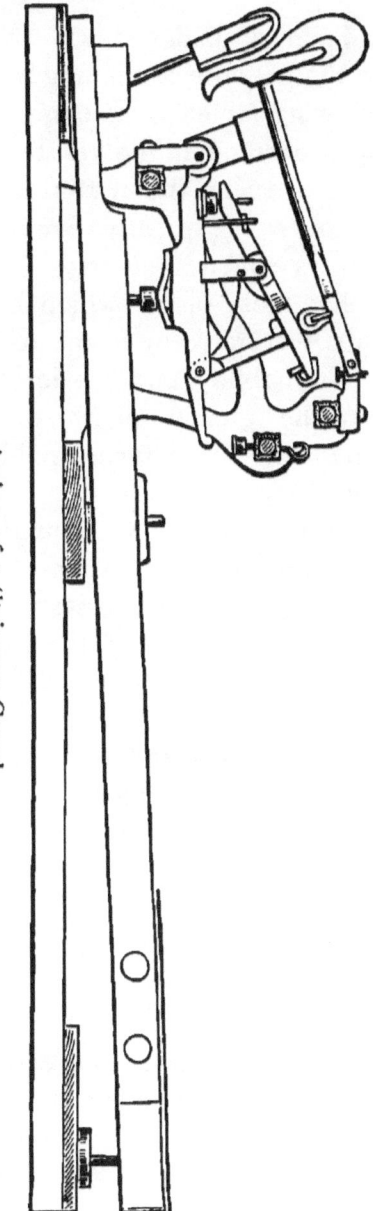

The key, which is disconnected from the action-frame, bears an upright knob called an action-pilot. This knob pushes up the lowest lever of the action, which is pivoted at the left. When it rises to a certain point this lever comes in contact with a button at the right. As the pressure from the key continues, the jack, which is under the felt knob upon the hammer-shank, begins to move beneath this knob; the pressure from the jack has already caused the hammer to strike the string. when the lever at the top of the action, which contains a collar for the jack, and is known as the "balance-lever," comes in contact with a small button behind the hammer-shank. It thereupon tilts up, receives the knob on the hammer-shank (called the walze) upon its edge, and allows the jack to escape from the hammer. The moment the key rises even a fraction of a centimeter, the walze slips back upon the head of the jack, and is ready for a repetition of the blow.

From Steinway & Sons' Catalogue, by permission.

Theodore Steinway believed, so exactly that the key
answers to the pressure of the finger down to the very
bottom, allowing for effect of dampness and no more.
If the finger overcomes the resistance of the key in
its descent too easily, the resources of the pianist in
variety of touch are greatly diminished, and legato
playing always suffers, and with it nobility of style.
It is here that those pianos which have a light touch
as their chief feature seem to me to fail. Lace-work
flourishes, but nobility and passion fade for lack of
elasticity. The greater depth of touch prevalent
among American pianos favors nobility of style, but
not execution.

Although a large majority of builders buy their actions ready made, to save expense, the great houses
make their own. Each possesses its own special features, usually the product of the experience of generations. This simple little contrivance—a pair of
levers with an escapement—embodies, in its present
perfection, the world's toil for two hundred years.
This is no mere wooden thing. It means the yearn-
ing and striving which are the moving springs of
human life,—the lifting powers of civilization. The
unspeakable awe and wishfulness which come from
consciousness of our inner and, as we think, immortal life have brought it to pass.

"Man would not be the finest creature on the earth
if he were not too fine for the earth," said Goethe.
The first step in civilization is a divine discontent with
mere living,—mere animal enjoyment of the fruits of
the earth,—the instant voice whereof is music. This

discontent with material life — this sense of our divine birthright that language cannot articulate — has set civilization the problem: "Contrive for me a mechanism by which I can first translate, then idealize into music, every feeling of humanity. God has given each man one voice. Make, O Civilization, each of my ten fingers into a voice. What the Almighty has done in the delicate muscles and levers of my arm, do thou duplicate behind the keys in springs and levers."

Such is this action, its history recorded in patents stretching back through centuries, and covering again and again every curve, every angle, every grain of weight, every millimeter of length and breadth; taking up the lifetime of generations of men who could not understand each other's speech, but whose thought met in something higher than speech. Future races will scrutinize its adjustment, so delicate that its manufacture has long passed out of mechanics into art, and read therein the witness of our higher nature, which has pressed forward this passionate labor of invention, to give you and me new power to express how we feel and thereby to set a new sign of immortal life before us.

In the hour when I saw this, the walls of the factory wherein I stood stretched upward to the grandeur of God's temple; and the wrinkled face of the workman beside me, his eyes resting lovingly and proudly on the beauties of the action before us, became glorified in a priesthood whose majesty he knew not. It is the wonder and pathos of life that they who serve its deepest mysteries — yes, even the holy of holies — have no significant initiation, no outward badge. Their badge is but toil's superscription in the lines of face and form; their initiation but the long discipline of faithful labor. Theirs is but a matter of regulating a few springs and levers, but the levers are among those which lift humanity.

THE SOUNDBOARD.

PLACE before you this glass plate, supported, as you see, upon a standard. Chladni laid the foundations of acoustics when watching the motion of sand strewn on such a plate as this. I draw my bow across its edge, and produce an audible tone; the sand arranges itself on the surface of the plate in geometrical figures, and you guess at once that the heaps you see forming gather upon the quiet portions of the board,—that is, upon nodes,—and that the vacant spaces discover vibrating segments, from which the sand has been tossed away. Such a plate, subject, according to its points of support and attack, to changeable nodes, representing different partial tones, is the soundboard of a piano. I strike this tuning-fork; you do not hear it. But now, when I bring its stem in contact with the wooden table, it is clearly audible. The table is a soundboard—that is, a surface capable of reinforcing the vibrations of a sounding body. It is also capable of sustaining its own independent system of vibrations.

The soundboard of a piano consists of three parts: the board itself, its ribs, and its bridge. The bridge supports the strings. When a string is in place upon a piano, it is attached at one end by its wrest-pins to the wrest-plank; at the other by its hitch-pins to the string-block. The pull of the strings is counteracted

Chladni's Glass Plate.

by the tension-bars of the metal frame. These bars run above the strings from hitch-pin plate to wrest-plank, and hold the two apart. On the soundboard, near the back, before the hitch-pins, stands the bridge; in front of the wrest-pins a flange of the wrest-plank plate projects under the string, and practically forms a second bridge. A series of pins on the upper side of the soundboard-bridge hold the strings in place. It is

10

the office of the bridge to transmit the vibrations of the strings to the soundboard. Its curve defines the vibrating length of the strings, and its height determines the amount of downward pressure which they exert upon the soundboard.

The bridge of the piano that we are studying is made of alternate veneers of hard and soft wood, because such veneers are better conductors of sound-waves than solid blocks of wood. Theodore Steinway found that the vibrations of the strings run from one end of such a bridge to the other, and set in vibration all the fibers of the soundboard beneath. This composite bridge, like the iron frame and wooden bracing of which I shall presently speak, was perfected only a few years before the death of its inventor.

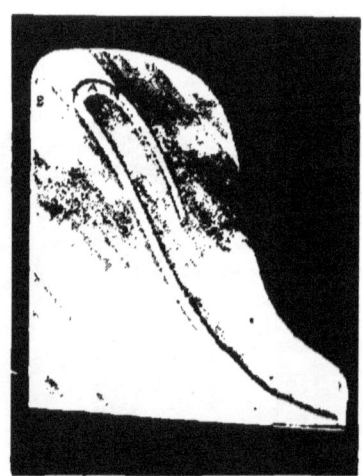

Soundboard, showing Ring-bridge (upper side).

The three represent the incessant study of his life. Sound is propagated through different kinds of wood, and in different directions through the same block of wood, with widely differing velocities. It was in scientific investigation of such phenomena, and in practical application of the laws thereby discovered, that the genius of this great artist found its congenial employment. Solid maple or beech bridges, however, are still in common use among the majority of makers. Since the fan-scale made its appearance, the bridge has been brought as much as possible into the center of the soundboard. It should be as nearly parallel as possible with the fibers of the board.

The pressure of the strings on the soundboard must be met by support from beneath. This is usually attempted by the use of ribs, but their application varies in theory and practice. Too many ribs stiffen the board and stop its vibration; too few allow it to play up and down, and thus refuse the vibrations of the strings. If the pressure of the strings is not counteracted, the soundboard sags; then the strings lie loose on the bridge, and do not communicate their vibrations to it, whereupon the tone

Soundboard, showing Ribs and Tone Pulsator (lower side).

becomes weak and thin, like the tinkle of a music-box. During the last century every imaginable plan has been tried to sustain the soundboard without spoiling its tone. Nothing better than good ribbing has been devised. Every piano-maker has his own experience and secrets concerning the shape, bracing, thickness, and attachment of this exquisitely sen-

sitive part of the piano.   Its treatment largely marks
the school of building to which the maker adheres.

Blüthner holds that the tone is not affected by the
direction which the annual rings of the wood hold toward the strings; but the rings must form straight lines.
Fine-grained wood, according to him, is better for the
treble side; coarser for the bass; the space between to
be built up gradually, so that unlike pieces do not
come together. During the last century the thickness
of the soundboard has increased with the size of the
piano—the treble perhaps a little thinner than the
bass, custom diverging. Blüthner proposes making
each soundboard out of the wood of a single tree. Coni-
fers are supposed to satisfy acoustic requirements best.
Pine, larch, cedar, and mahogany have been tried,
with corresponding modifications of tone, but spruce
is the usual choice. Distinction is made as to place
of growth. Spruce from cool, stony uplands is of finer
grain than that from warm bottom-lands; boards from
the north side of a tree than those from its south side.
This difference between the two sides of the same tree
appears to have impressed makers of other musical
instruments, for Duborg mentions south-side wood as
better for violins. But the vast manufactories of to-
day forbid such primitive practices. After the sap
begins to run, the wood is unfit to cut. The resins
within the wood, like the varnish applied without,
greatly modify its tone.

Before the soundboard is set in the piano, its ribs
give its surface an upward curve, which resists the
downward pressure of the strings.

Rimbault quotes the following lines, copied from the soundboard of a harpsichord in his possession:

> I once was living in the woods,
>   But now I am cut downe
> By stroke of cruel axe indeed,
>   But yet to my renowne:
> For while I lived I spoke naught else
>   But what the boistrous winde
> Compelled my murmuring strains unto.
>   But being dead, I please ye minde
> And eares of such as hear me singe,
>   So pleasant is my music's ringe.

THE WOODEN BRACING.

WE have seen the evolution of soundboard and strings coincident with the entire development of our civilization. The wood and iron bracing which is the formative element of the modern piano is the work of the nineteenth century. When the President of the New-York Piano-Makers' Association laid before its members, at a late annual dinner, the initial story of our metal frame, he threw an unexpected light upon the meaning of his art. It was the expulsion of sixty thousand Protestants from Austria during the first half of the eighteenth century that sent the founders of our iron industries to America. So something stronger than iron, and infinitely nobler than mere mechanical skill,— even religious conviction,—is wrought into the tensile strength of our American piano. For the descendants of these same exiles, by perfecting the science of iron-casting, made old Alpheus Babcock's "whole iron plate" first successful in America. This iron frame, the work of American genius, constitutes the creative feature of the piano of to-day. The speech from which I have quoted adds another essential fact. Just as Italian and Tyrolese forests rendered Amati violins possible in Cremona, so American lumber has carried piano-making to its highest perfection here on our Atlantic seaboard.

To understand the necessity for the iron plate, let us look at the wooden bracing.

At the back of an upright, and at the bottom of a grand piano are certain beams of wood corresponding to the beams and rafters of a house. Upon them everything rests; into them everything is bolted. The framing is the security of the tuning; the slightest yielding at any point throws the whole scale into disorder. No part of a piano may be loose enough to jar, or tremble in its place. Every part must be so bound to every other part that there is but one absolute whole, just as the flesh and bones of a human being are grown together into a whole.

The traditional build of this wooden frame has shown great changes. In old times a number of timbers ran lengthwise through the body of the piano, with cross-braces dovetailed in at right angles. The whole was united by a sort of rim, made up of short pieces, that supported the edges of the soundboard.

The wooden wrest-plank, which bore the tuning-pins, was likewise bolted to this frame. The weight of this shapeless mass was enormous in proportion to its strength; for it was constantly warping and bending. Originally it offered the only resistance to the whole drawing force of the strings. Since the wood alternately swelled and contracted in consequence of atmospheric changes, the strings were never under the same strain from their tuning-pins, and therefore never in tune.

It was in 1799 that Joseph Smith took out a patent for metal bracing, to strengthen the case enough "to admit of introducing a drum, tabor, or tambourine, with sticks or beaters, as well as a triangle, into the body of the instrument." Smith's English patent was followed the next year by a plan for metal framing, by Isaak Hawkins of Philadelphia. England made the second step when James Shudi Broadwood, in 1808, applied three steel tension-bars above the strings to prevent the treble part of his grand piano from flatting. In 1820, Thom and Allen, two workmen employed by Robert Stodart, invented a system of bracing with hollow metal tension-bars applied over the strings of grand pianos. These bars, firm at one end, were fastened at the other in a movable slide, and yielded with the expansion and contraction of the strings. The metal wrest-plank plate came later.

Babcock's "whole iron plate"—the complete frame, tension-bars, and string-plate cast in one piece—was patented in 1825. This model, passing through various hands, was much improved by the Chickerings (grand

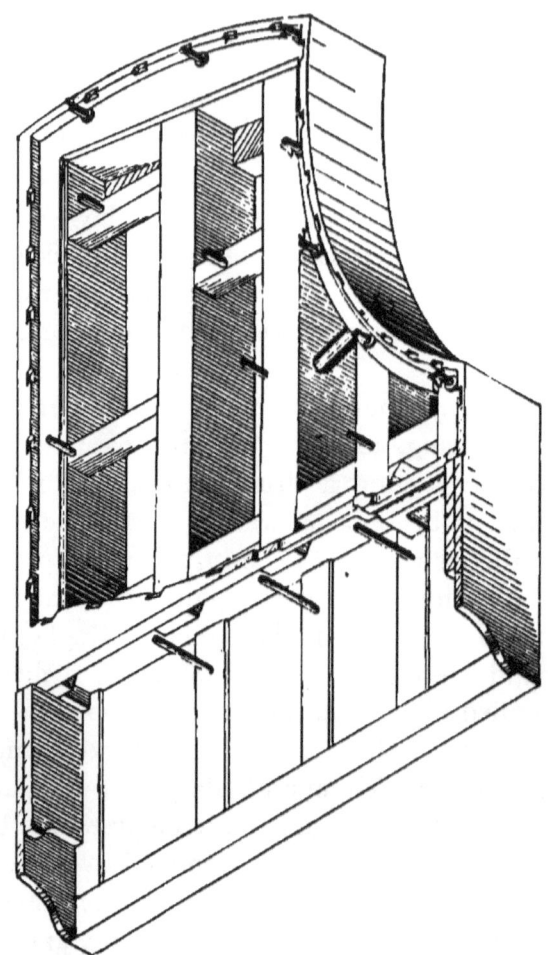

The Usual Method of Wooden Bracing (from a Chickering Patent).

plate with agraffes) in 1843. But the practical relations of wood and iron to each other were still little understood when, in the middle of our century, Jonas Chickering being dead, Theodore Steinway came forward to

take his place in history. The experiments made by
the latter required the capital of a rich and powerful
industry. The Steinways began with only ten thousand
dollars, and for years every day saw them at the
bench; but as soon as circumstances made it possible,
Mr. Steinway carried out the dream of his life, and set
his piano upon a scientific basis.

In a modern Steinway grand piano the strings pull
from the hitch-pins at one end toward the wrest-pins
at the other, with a force of from twenty to thirty tons.
The wooden braces, and iron tension-bars, running
nearly parallel with the strings, prevent the piano from
collapsing endwise; but a cross-brace is required to
strengthen the wrest-plank. M. Bord of Paris met this
need by the invention of a transverse bar, which he
called a *capo d'astro*, from the well-known contrivance applied to the strings of a guitar. This bar, for
which M. Bord intended no acoustic application, is the
nucleus of Theodore Steinway's *capo d'astro* bar in the
frame before us, which not only forms the upward bearing for the treble strings,—pressing them upon the
front of the flange beneath, made by the extension of
the wrest-plank plate,—but provides the node of separation for the duplex scale.

The wrest-plank plate, the transverse bar, the longitudinal bars, and the hitch-pin plate make up the
various parts of the iron frame,—cast in the piano
before us in one piece.

My three pianos show the three stages of upper
bracing: the Broadwood, without brace or iron, its
corkscrew strings (hooked by eyes over the hitch-pins)

hardly strained to the tension of a violin-string; the
Nunns & Clark, with its tiny plate at the right; and
the cupola frame of a modern Steinway grand.

Upon this frame you will see, at the wrest-plank and
hitch-pin ends, little rib-like protuberances upon which
the strings lie. Another plan is to screw or cast in
the plate eyed pegs, called (by Erard, their inventor)

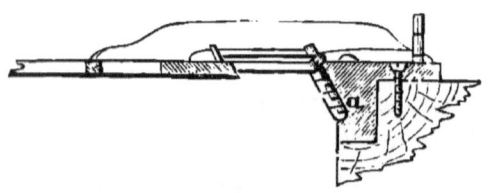

a, Old-fashioned Agraffe.

agraffes, through which the strings are threaded; or
the strings may lie on a rib on the plate.

Looking at my frame again, you see that the hitch-
pin plate does not bear the groups of pins at the same
elevation. The differences represent the different
heights at which the bass, middle, and treble webs
of strings cross each other. The hitch-pin plate itself
curves upward from the flange that forms its edge.
The upper surface of the curve is marked by a series
of open rosettes, which remind you of the rose of a
guitar, and answer the same purpose. This arched
frame, with its braces, is the famous "cupola" which,
like the fan-scale, revolutionized piano-making.

But the principle upon which the wooden braces of
the soundboard are arranged is still more original and
daring—if a work of evolution that can be traced

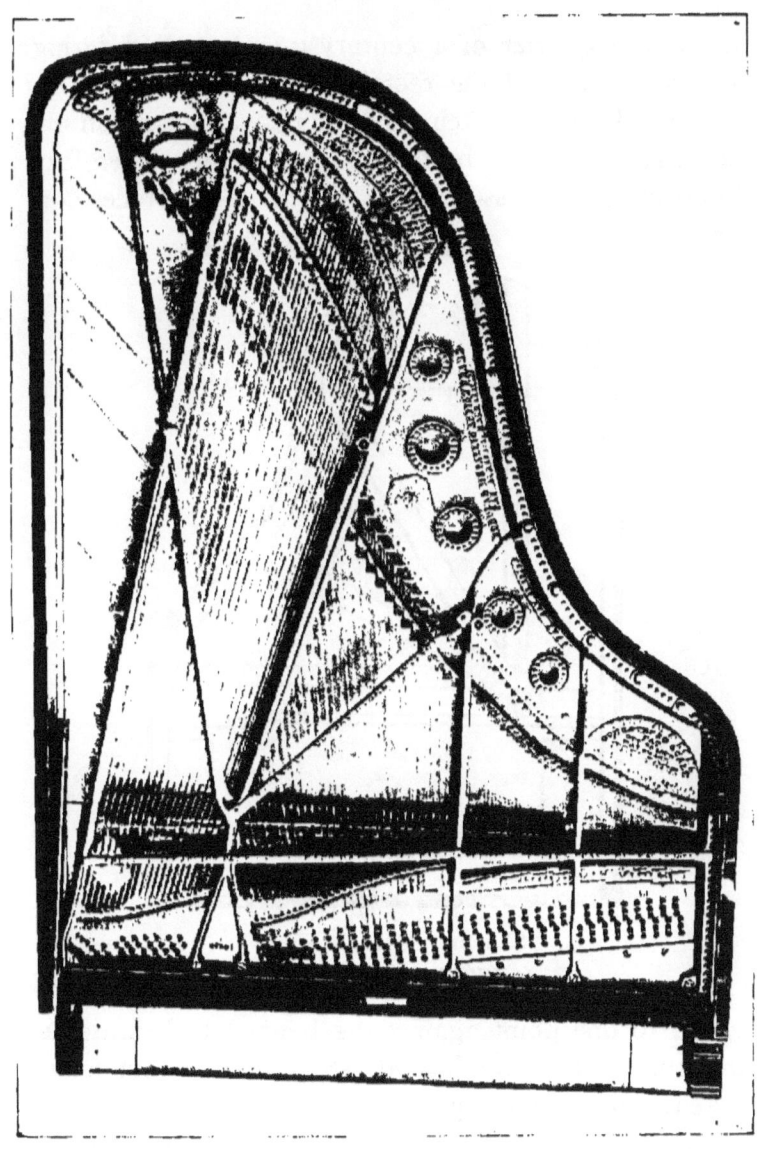

The Stringing and Iron Cupola Frame of a Steinway Grand,
the Soundboard in place.

through a quarter of a century may be called daring.
I have spoken of the rectangular bracing of the old
piano. Theodore Steinway's plan was based on the
fact that wood-fiber best conducts vibration when least
interrupted by cross-contact. He made his braces con-

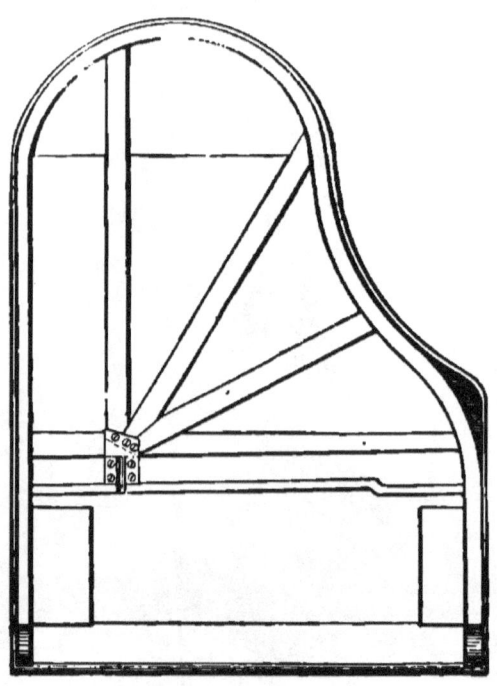

Showing " Metal Shoe."
Permission of Steinway & Sons.

verge to one point against the beam which underlies
the front of the soundboard, where they are supported
by each other and by the beam, in combination with
a metal shoe which bears against the iron frame.
Their outer ends abut against a rim of veneered wood

—the fibers sometimes twenty-three feet long—which forms the inner case of the piano. The front of this rim lies above the cross-bar, so that the edge of the soundboard is supported by it throughout. The rim vibrates powerfully beneath this edge, and redistributes the vibrations of each portion of the board throughout its entire circle. The radiating braces, set like spokes against the rim of a wheel, not only offer great resistance to the inward pull of the strings, but, since they only abut against the rim, without interrupting its fibers, they allow it to vibrate freely. This rim, like the outer case, consists of many veneers of wood pressed and glued into the required curves. Upon the rim, or what answers to it in other pianos, is glued the soundboard; above the soundboard, and bolted through it to the rim, is the arched iron frame. Every precaution is taken to connect the vibrations of the wooden parts of the instrument, and to separate them from the iron. To the exquisite skill with which this is effected, and to the perfect workmanship and finish of each several part, may be traced the wonderful resonance of the Steinway piano.

The outer case of a piano, when well managed, has much more to do with the tone than people imagine. It vibrates powerfully with the sounding parts of the instrument, and, as it is in contact with the outer air, impresses its motion directly upon it. It happens then that the plainest cases often lend themselves to the best tone. But the delight of turning musical instruments into jewels has always been active.

The museum of the Conservatory of Music at Paris

contains an Italian spinet, made by Annibale Rossi, in
1577. The case (I translate from Blondel) "is covered
with panels and borders of ebony, richly decorated
with plaques of lapis lazuli and precious stones, which
are framed with cartouches of ivory, finely and delicately carved. Each panel is itself surrounded with
ornaments of ivory, incrusted with rubies, topazes,
emeralds, and fine pearls. The panel of the keyboard
is ornamented with macarons and arabesques alternately. On the transverse bar, which is also incrusted
with fine pearls, are placed three graceful figures, in
ivory, of Amors playing the viol. The white keys are
made of agates, variously framed in ivory; the black
of lapis lazuli. The keyboard is terminated at each
end by consoles, decorated with very elegant figurines
carved in boxwood." France has always longed to
turn the piano into an article of "vertu," and Pape
long ago constructed one in a case covered with ivory,
for the Duchesse de Berri.

Erard followed (the fervor of the description would
be lost in translation) with a piano " en pur style Louis
XVI., c'est revêtu du thuya blond, manteau luxueux,
qui se marie admirablement avec les dorures dont il
est chamarré, et qu'enrichissent encore de délicieuse
peintures d'une finesse coquette et mondaine, dues au
pinceau de Gonzales." Doubtless the piano was excellent, but there is something in the innocent, but Philis-
tine glee of the above lines that rankles in the soul of
the lover of piano-making. The spirit of that abomi-
nable story of Salvator Rosa, that has made him the
tutelary genius of too many piano-makers, lurks in it.

Top view of Theodore Steinway's wooden bracing. The cut
shows the bent rim, and the wooden wrest-plank, which
underlies the wrest-plank plate. (Without closing rim.)

When told that his miserable clavecin was not worth three écus, "Then I 'll make it worth three thousand," cried he; and forthwith painted a landscape on it. But beautiful pianos exist that are in keeping with the reverent spirit of music—the feeling Moscheles expressed when he quoted, about his Erard, Schiller's line:

In dem schönen Körper mus auch eine schöne Seele wohnen.

This lecture has dealt only with such broad outlines and well-defined methods as belong rather to science than to art. Even these outlines are incomplete. Founded on that "observed order of events," which Huxley defines as scientific law, must rise a subtler craft, or piano-making would be but a matter of engineering and mechanics. This nobler technique, residing in the secret tradition of the house, or even in the personal gift of the artist,—this art that touches every detail, and from the commonest use or material brings forth the surprise of beauty, is the characteristic charm of piano-making.

It is its threefold character of art, science, and handicraft that makes at once its glory and its danger. No other industry combines in itself such diverse and contradictory elements, or unites in its conduct so many phases of human life, so many opposing types of character. None other demands such culture, principle, and nobility in its prosecution.

As I have worked on this lecture, the poetical side of the building of instruments has been borne in upon me. Pat to my thought came a little story in " The Century Magazine."

Rack, a darky, and his naturalist master, noticed

The same wooden frame, seen from beneath, exposing
bent rim and wooden bottom.

and disputed over three facts: a large opossum, a
board constantly drummed upon by a woodpecker,
and a knotted bough where a mocking-bird daily
brought his mulberries to squeeze and sip their purple
juice. The master held no scientific correlation of
these objects possible; but the darky differed. Six
years did the dusky inventor hunt the opossum, and
finally caught him, to make his skin the drum of a wonderful banjo, the rim of which was cut from the woodpecker's board, and its head and neck from the bough
stained purple with the mocking-bird's berries. It was
a wonderful instrument. When it was done, master and
man, mad with its sweet sounds, sang and danced to
it all night long. On the rim was carved the legend,
"Dis am de coriolation." Now every work of art is
equally a "coriolation," and like this banjo has intoxi-
cation and magic in it. It has been easy for the author
of this very intuitive sketch to fuse his poetical elements
into a whole. The artistic creation of the banjo is evi-
dent, and its poetical aspect at once moves us re-
sponsively. It was planned and made by one single
individual, and through it he expressed himself. In a
piano, made in hundreds of pieces, passing through
hundreds of hands, it is more difficult to discover the
poetry, since personality seems absent. Similarly the
history of nations, involving simply a chronicle of wars
and famines, is dry compared with the novel dealing
with one living, loving soul. But when it is seen that
nations, like individuals, have a birth, development,
story of joy, heroism, sorrow, patience, and suffering,
history becomes the romance of giants.

Such is the story of the creation of the piano: not like the idyllic chronicle of banjo and flute; not like the quaint medieval tale of the violin; but a biography whose dimensions coincide with those of an era of human development,—the history of a heroic struggle,— the manifestation of a divine spirit striving ceaselessly for perfection in this its mortal form.

III.

THE ARTISTS OF PIANO-MAKING.

"By this art, which whoso will may sacrilegiously degrade into a handicraft," adds Teufelsdröckh, "have I thenceforth abidden."
Carlyle.

"ART," says Emerson, "is the path of a creator to his work." Not all creations of genius are art, but all have one property of art in them,— they discover and work out under nature's laws. Whoever divines the hidden laws of nature, and is able to work conformably with them, brings to sight some new thing that, by the personal operation of the man's own spirit, which has brooded over it and united itself with nature to produce it, is stamped with his personality. Such a production we call a creation of genius. But there is something more than creation merely in art, else would the telephone or the engine be the highest art, for they are the working out of an idea. Art has for its soul not an idea merely, but an ideal. It is not enough that truth, and therefore law, should be uttered materially. These are things that in their simplest manifestations promulgate their own edicts, in decay, and fire, and gravity. It must be more than personal combination of truths and their manifestation; it must be a revelation of that particular kind of truth that touches us with a sense of what we call beauty. Strictly speaking, the ideal is an idea that has beauty working in it, organizing it, holding it together, as it were. Beauty is a hard thing to define. The day "God saw everything that he had made, and behold it was very good," marked its first appearance on earth. What if we should say that a sense of beauty is the responsive thrill of our divine nature, as God here and there recognizes his own handiwork? Susceptibility to beauty as the revelation of an ideal is one of the properties of our divine nature. As humanity rises, it increases; as it deteriorates, it disappears. A South Sea Islander who has just killed and eaten his wife, is probably not appealed to by beauty.

So now we may carry our exposition a step farther, and say that a work of art is generated by an ideal of beauty in the soul of its creator, to which he gives material form, working conformably to more or less obscure natural laws which he has the gift to divine and to follow. And as the divining and following are the actions of the whole man, his creation is at every point qualified by his own nature, and comes out in the image and under the superscription thereof. And so it is the expression of law, of ideal beauty, and of the man himself.

The *sine qua non* of art is ideality, as the *sine qua non* of man is the soul, of which ideality is a property.

Khaled the Genius, according to Marion Crawford, was promised a soul if he could win the love of his human wife. He had a human body, and personality, and intellect; but for many months he could win from Zehowah good-fellowship, but not love. At last his companionship became dear to her, and once, when his life was in jeopardy, she said, " Khaled, I love thee." Then the angel brought Khaled his soul, and it entered into him and became part of him. And Zehowah looked into his eyes and saw the living soul flaming within. " If you had always been as you are now, I should always have loved you," said she. And so it is with art.

The expression of the ideal the artist knows to be godlike, and the longing for such expression, over-masters all else. The expression of the man is unconscious, and the instant the artist seeks it he has lost sight of his ideal, and his work is no more true art; but were the personal flavor lacking, it would equally fall short of art. If this definition be allowed, then piano-making is true art; and I hope to show you that it springs from an ideal of beauty; that it is a revelation of law; that each instrument embodies the personality of its creator.

We cannot think of an art as springing suddenly and sporadically into existence; it has a lineage and heraldry of its own. It quarters the arms of heroes. An art implies its national artists, each with his train of students and his characteristic school.

Piano-making is a new art of Christendom; the whole knowledge of musical-instrument-making has gone into it, either in mechanism or ideal. Violin, harp, harpsichord, and organ have each contributed something; and every phase of our civilization, and every nation which brought that civilization forward, has lent a hand. So the piano is composite, like Christendom, and Christianity itself, else it would not represent our age.

If we seek the name that more than any other is so filled with memories of art, beauty, vice, magnificence, statesmanship, letters, that it brings the whole Renaissance before us with a word, that word would be Medici. The nature of the piano seems to be suddenly clear to us when we know that it was at the court of the Medici that Bartolomeo Cristofori made the first piano of its race. The name of Cristofori appears in Hart's book as that of a violin-maker. He was a contemporary of Guarnèrius del Gesù, and of Antonio Stradivarius; and he was not only a maker of violins, but of harpsichords. At the beginning of the eighteenth century the noble houses of Europe, which rivaled each other in the patronage of men like Stradivarius, gave a stimulus to artistic production. Of Cristofori himself I have no particulars; but Stradivarius, who worked day by day for three quarters of a century in his white skin apron, and nightcap, belonged to the intelligent middle class. He owned his house, wrote his labels in good Latin, held political opinions, and dealt with crowned heads. Cristofori, belonging to the same class, must have cherished similar habits, principles, and ideas.

These were the flower; the root was far back, when artisans combined into guilds, like that of St. Luke's of Antwerp. The principal rules of the instrument-makers when they entered this guild in 1559, quoted by Fleming, are as follows:

"I. A masterpiece to be made and exhibited by each incoming member.

"II. The examination and approval of the master-piece on oath by an examining committee of the craft.

"III. The signature by the maker of all the work sold by him."

Here were conditions most salutary and stimulating to individual talent; and Antwerp was the queen of harpsichord-makers for a century and a half. Ruckers, her greatest maker, is still represented by his instruments, exquisitely finished, and playable to-day.

The Netherland school has limned the Guilders in immortal colors. We can carry away a memory of them and of their guilds, as we see what their genius bequeathed to the piano, which inherits the shape, the general structure, and the mass of scientific tradition accumulated by harpsichord-makers. Nothing is more dainty than the ivory carvings, the case inlaid or painted by great artists, the graceful outline and proportions, of a harpsichord. "Father," said a poor artisan's child, pausing to look at the one in the Metropolitan Museum, "is that an angel piano?"

But Italy, which, about the middle of the sixteenth century, began to perfect the violin, was to give the first permanent form to our modern piano. Italy, the lover of tone, in accomplishing the violin, taught the ear to demand what no keyed instrument had ever given, and, following her genius, she passed beyond

spinet and harpsichord and invented the hammer-
clavier. Her knowledge of woods and varnishes,
and of the manipulation of musical instruments, stood
her in good stead. The piano made by Cristofori
may be successfully played to-
day, and our modern instrument in its several parts has
been developed consistently
from the germs created by his
genius. The same instinct for
form and knowledge of materials that enabled Stradivarius
and Bergonzi to measure the
backs and bellies of their fiddles, to carve the heads so that
the beauty apparent in every
curve was but a revelation
of the ideal of tone that demanded this and no other
form, were busy in Cristofori
when he built the first piano
and invented its action.

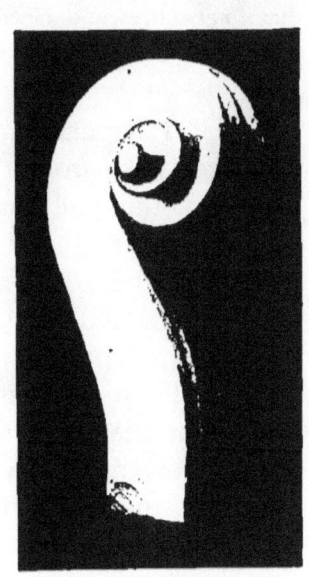

A " Strad." Head.
To be grafted or spliced on.

The following extract from the account of Scipione
Maffei states the artistic position of the new invention:

Of the diversity and alteration of tone in which instruments
played by the bow especially excel, the harpsichord is entirely
deprived, and it would have seemed a vain endeavor to make it
so that it should participate in this power. . . . Signor Bartolomeo Cristofori, harpsichord-player, has already made three [ham-
mer-claviers] of the usual size of other harpsichords, and they have
all succeeded to perfection. The production of greater or less
sound depends upon the degree of power with which the player

13

presses on the keys, by regulating which not only the piano and forte are heard, but also the gradations and diversity of power, as in a violoncello. . . . To some it appears that the tone is more soft and less distinct than the tone of ordinary ones, but in a short time the ear so adapts itself and becomes charmed with it that it never tires, and the common harpsichord no longer pleases; and we must add that it sounds yet more sweet at some distance.

But if Italy demanded a pure, fluent, and musical tone, free alike from the tick of the tangent and the scratch of the quill, the genius of Germany was intent on very different ends. The Gothic instinct is for wealth, richness of ornament, fullness of energy, of beauty, of sound, of battle; not limpidity and grace, but power and passion; passion rampant in Wagner, and chastened into sentiment in La Motte-Fouqué. Therefore, from the moment Silbermann brought forward the new invention, the constant struggle of Germany has been to endow it with the properties for which she longed most intensely. Germany, working with America, may be said to have given the piano its supreme charm—its present fullness and quality of tone. The contribution of Austria was very different. Lightness of action, ease, fluency, a complacent and agreeable tinkle, were needed for execution, and Stein to make and Czerny to teach arose obedient to the genius of the nation; and, having achieved its work, the school of art which they represented perpetuated its own special ideals at home, and gave its best to the general ideal of Christendom. Mozart admires Stein's pianos, "because I can lift the finger or leave it on the note, for the sound is not prolonged

beyond the instant in which it is heard. I strike the
chords as I please; the tone is always the same. It
never shivers, or fails to sound, as is the case so often
in other pianos. Stein labors not for his pecuniary interest, but for that of art. He says, 'If I were not
myself a passionate amateur in music, my patience
would long ago have failed me.' A tone which
vanished as soon as heard, and a touch that could not
be modified by the player, would be unpardonable
defects in a modern piano. But the first defines Stein's
success with his damping, and the new certainty of
stroke offered by the second generated the virtuosity
of to-day. For the evolution of the piano was the
evolution of an ideal. It mattered little whether Stein
and Streicher handed down their own special action
and tradition : both were faulty in the extreme. What
availed was the conception of a touch that responded
to the slightest effort of the player. The literature of
music that followed its appearance at once crystallized
the ideal imperishably. And since music that lives at
all always transcends its source, this literature called
unceasingly for the realization of what the Viennese
school barely promised, and necessitated the work of
Erard and the Parisian school that followed,— the tone
that reflected, like a mirror, the slightest alteration of
the player's mood. At the same time England was
erecting her own school, different from all the others,
and for a while their leader.

No sooner had Silbermann succeeded in reproduc-
ing the plans of Cristofori than Germany turned from
the harpsichord model to her other keyed instrument,

the clavichord, and Zumpè, emigrating to London, began to make square pianos with an action clearly derived therefrom. It was not as good an action as
Cristofori's, and the tone of the piano was far less
musical and limpid; but the rich middle class of England offered a ready market, and, once launched, improvements came fast.

Among the men who founded the English school,
conspicuous shone the Broadwoods, Clementi (Clementi & Collard), and Cramer (Chappell & Co.): two
of these names immortal for contributions to the literature of the instrument. The English love of good,
firm building, good workmanship, lent special features
to the English piano. Its striking tone grew thick and
loud, though without much duration. It was to be
exported all over the world, therefore inquiries about
bracing and stringing were set on foot, and the skill of
English iron-workers fitly gave the first notion of metal
framing to England. London, the mart of civilization,
found easy access to the woods and ivories that came
from all over the discovered world. Her creation
gradually took on the characteristics of the English
piano we know. Still France clung to clavichord and
spinet, till presently Erard, working out in the direction
of French feeling and need of expression, produced his
repetition action. Here was the limpid, fluent tone of
Vienna, with power enough to set in vibration the
large instruments of England, and a firmness, delicacy,
and purity of touch altogether new, such as to allow at
once of taste and passion. Race characteristics are as
active in national musical tone as in dialect.

This national taste in tone-quality is illustrated by an experience of Moscheles. I quote the passage:

Moscheles wished, by using alternately a Graf and an English piano at one of his Vienna concerts, to bring out the good qualities of both ; and Graf, foreseeing the favorable issue of the contest to himself, generously labored to put Beethoven's Broadwood into better condition. "I tried," says Moscheles, "to show the value of the broad, full, though somewhat muffled tone of the Broadwood piano, but in vain. My Vienna public remained loyal to their countryman — the clear, ringing tones of the Graf were more pleasing to their ears."

Moscheles also supplies a chronicle of the evolution of the Erard piano. I quote from " Recent Music and Musicians." " I was the first to play upon one of the newly completed instruments (1825), and found it of priceless value for the repetition of the notes. In the matter of softness and fullness of tone, there is something yet to be desired."

In 1828. "Externally the instrument is all that can be wished, but the tone of the higher notes is still somewhat dry, and I find the touch still too heavy."

In 1830 he finds his progress as a composer greatly favored by the improvement in Erard's pianos, " their organ-like tone, and full, resonant sounds,—'a very violoncello,' he would say, praising the tone, which he could prolong *without the use of the pedal*" (apparently the first appearance of our modern singing tone).

Six years later, "H. Herz introduced his seven-octave piano in the London concert market, but the tone was declared to be thin."

In 1853, he writes that the new Erard "has the softness of a flute, and the power of an organ."

This "flute-like softness," which is so grateful to the
ears of the artist, marks the growing success of the
piano-maker in bringing his soundboard into vibration. The constant comparison with organ and violoncello shows the potent ideal of tone not yet realized,
but ever hovering before the fancy of the German pianist; for the Erard of 1853 had hardly passed the
threshold of the art of piano-making which we know
to-day.

Thalberg's enthusiastic description of Erard's piano,
which largely opened the way to the success of that
instrument, marks the precise point at which European
piano-making had arrived when, Chickering having patented his grand plate, the Steinways began their in-
novations. It is clear that America was already ahead
in matters of bracing.

"I am able to give an idea of the degree of perfection attained in the construction of the piano at the
present day" (1851), writes Thalberg of the new grand
piano exhibited by Erard in London during that year.
"This instrument is 8¼ feet in length, and 4½ in its
greatest width, being heavily braced with wood below
the strings, having a complete system of metallic bracing above the strings firmly abutted, and consisting of
longitudinal bars let into metal at each end, and having a curved side formed of a number of separate
pieces glued together in a mould to insure durability
and fixedness of form. Its sounding-board extends to
the frame on all sides, except the space left for the ac-

tion. The strings are made entirely of steel, and of wire so thick that the tension necessary to bring them to the proper pitch produces an aggregate strain of at least twelve tons weight, while they are passed through studs (agraffes) drilled into the metal wrest-plank, thus giving the strings an up-bearing position, which prevents the slightest displacement of the point of contact by any force of the hammers; and the system of placing the strings on the instrument, determined by accurate acoustic experiments, causes them to be struck by the hammer at the precise nodal point which produces the freest and most perfect tone. . . . The action is regulated to such perfection that if any note is missed in execution upon it, it is the fault of the player, and not of the instrument."

It was not by accident that France initiated the changes which made the fate of the piano as the favorite instrument of civilization certain. France was the last country of Europe to accept the piano —"that regular brass kettle," as Voltaire dubbed it. Her literature is full of squibs at the expense of its English coldness. As she accepted it from England, it was capable of neither sentiment nor piquancy. France, too, already possessed a national quality of musical tone,—that thin, sensitive soprano quality we recognize alike in her singers, her violinists, her organs, and her pianists.

She no sooner adopted the piano than this quality appeared in all her makers, and was carried to great perfection by Erard and Pleyel. It was well into the beginning of the present century that this change came; and America had already begun to make pianos for herself. Philadelphia, New-York, and Boston each possessed her own group of makers. Presently the Babcocks began to invent and experiment, and founded the

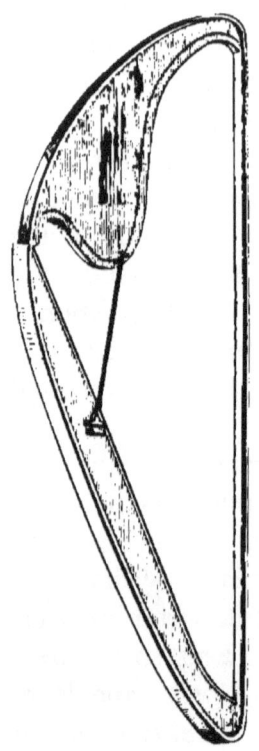

famous Boston school that ultimately, by its iron frames, altered the whole art of piano-making.

Because Boston is near home, or because the steps are really very well marked, the rise of a national school of art is extremely clear. There was Benjamin Crehore, who began to make pianos at Milton, Massachusetts, before 1800. He had three workmen— Osborne, and Alpheus and Lewis Babcock. Osborne owned a shop of his own presently, in which studied Dwight, the Gilberts, and Jonas Chickering.

Dwight invented the first factor in the American iron frame, the longitudinal iron bar; Alpheus Babcock thereupon invented the famous full iron plate, cast in one piece; and Jonas Chickering carried it on to the perfection of his grand plate. It must not be forgotten that what these men worked in was tone; bracing and scale are to tone what pigments and brushes are to color.

Alpheus Babcock's "Whole Iron Plate."

The iron plate answered the mechanical necessity which must be met in the effort to enlarge and qualify the tone of the piano that Boston was to carry forward. The plate of Chickering was the culmination of the invention, as Chickering was the flower of the Boston school. The joint efforts of the men who elaborated their common ideal resulted in a distinct national creation, unlike and superior to its predecessors. These men were all Americans; such traditions as they possessed were English. Working in Boston, they could not have been very much in contact with European piano-making. Their patents of actions, frames, braces, stringing, and the like, are very numerous, and show how independently and creatively they all studied, and how they all went on together, each carrying forward their common art. When the Chickering piano was perfected, it was a pure soprano. Emerson speaks of "temperament without a tongue." He errs, for music is its tongue; and the Puritan Bostonians, by force of their nervous, eager American temperament, evolved

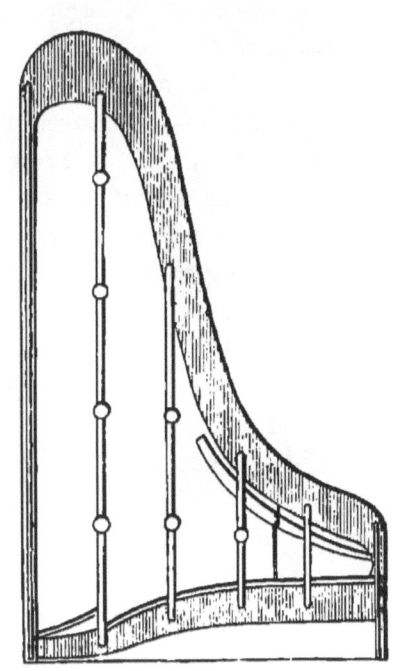

Jonas Chickering's "Grand Plate."

a piano of a quality akin to that of the French school. Temperament works immediately in all art; it is one of its essentials, and an art that crossed the temperament of its nation would be an anomaly.

But taste is a different matter: the great artist does not embody the taste of his countrymen. He leads it. The light action and thin, sensitive, but musical tone of the Chickering of 1870 belong to a symmetrical and consistent art work, of which America may well be proud. Thin as the volume of sound seems to us, it was enormous compared to its predecessors, and the scale was remarkably even from bottom to top. This was a great achievement, for the considerations that govern the seven octaves of the piano vary much as the maker proceeds upward; the struggle being first to obtain a clear, pure bass, and a full, resonant treble; next, to hide the places where the means adopted to produce one effect give place to those necessary for another.

The piano, too, is an extremely complex affair. Not one smallest item from casters to cover is without effect on its tone; and a piano, as a work of art that corresponds to an ideal of tone, is one in which every detail has been calculated and proved by experiment again and again. From the time of Crehore to the Centennial Exposition, when the Boston school was in its glory, there was waged a ceaseless upward struggle of nearly a century.

What a self-respecting, noble story the unfolding of our art has been! How charming the familiar pictures, as they pass through fancy and memory!

A stretch of yellow sand against the multitudinous sea: Hermes supporting himself on one knee, while he is looking at and twanging the fibers of the tortoise-shell he has just picked up. He is so absorbed in the ecstasy of invention that he fancies himself alone. But Venus, floating idly on the wave, is watching him, half in interest, half in mockery; while behind him, Apollo, bow in hand, looks intently over Hermes' shoulder, and Pallas by his side glances up to warn Venus to silence.

Guido Aretino, in his monastery of Avellano, his cowl thrown back from his handsome head, kneeling before the crucifix and offering to God the keyboard, which he has just completed. Manuscripts scrawled with neumæ, and the vellums of his own revolutionary tracts on music, jostle the missal and the scourge.

The young guilder and his masterpiece in the hall of the guild at Antwerp, the heads of the trade gathered in their rich dresses to examine his work; his sweetheart trembling with pride and hope for the decision.

Then, Duke Ferdinand de' Medici, surrounded by his courtiers in the workshop of his instrument-maker. They inspect the new invention, which owes its existence to his liberality. Cristofori himself, in cap and apron, explaining the models; the Duke's finger on the key. Here hereditary intelligence and the culture of generations meet the eagerness of genius.

Next, old Sebastian Bach, handsome and reverend in his wig and ruffles, playing the piano Silbermann has at last perfected on the Italian model, and rewarding the maker for his twenty years of study by the praise that opened the world to his handiwork. And, later, Bach, the guest of Frederick the Great at Sans Souci, extemporizing on the harpsichords and on Silbermann's piano; Voltaire, perhaps, a mocking listener.

Then Mozart, young and debonair, visiting Stein at his house, and inspecting his soundboards seasoning in the yard. And, in the next generation, Beethoven listening to Nanette Stein, as she plays his sonatas on the piano which she has made for him in her own factory: the worry and fret smoothed out of the musician's face for the moment in the comfort of her sympathy.

Old Backers, in his ugly middle-class London home, calling Broadwood to his death-bed, and confiding to his old fellow-workman the precious legacy of his incomplete grand piano, the labor of his best years, and begging him to carry it on to perfection.

John Broadwood, with his scientific friends, experimenting with the strings and weights of a monochord, calculating thereby the tension of the scale.

We can fancy James Broadwood going in his coach to call on Mr. Michael Haydn, and finding Cramer and Clementi there. Haydn descants on the revolutionary scheme of one Southwell, an obscure Irishman of Dublin, who will enlarge the keyboard to six octaves. Broadwood resolves to anticipate him, but replies dryly that "changes so extensive should await a pronounced public demand; one lives on pianos manufactured for the laity, not for artists." "Which sometimes induces artists to make their own," cries Haydn, offering snuff to Clementi.

Southwell, poor and discouraged, his Irish heat made patient by sheer genius, toiling over his upright piano in grimy London, patronized by the mighty Broadwood, who thrives thereby.

Erard, young, penniless, and talented, taking the commission that is too hard for his employer; and anon, his nephew, rich in his uncle's prosperity, pleading his cause among the nobles of France, and receiving his patent. Marie Antoinette owned one of Erard's pianos, made at her royal command; perhaps her kindly interest may be traceable in the successful suit.

Pleyel at Strasburg, under guard at midnight, composing literally for dear life the music that is to be a test of his loyalty, and thereby save or lose his head.

And Monsieur Henry Pape, seated in his private sanctum, surrounded by models of down-striking actions and reversed soundboards, and other nightmares of the art, lost in contemplation of the nap of his own hat, as he conceives the notion of the first felt hammer. And suddenly the scene shifts to Boston, and Alpheus Babcock, in small-clothes, his smart apprentices larking at his heels, goes decorously to meeting on the Sabbath day, and between the sixteenthlies turns over the measurements of that famous iron plate which unlocks the gate to our piano as we know it to-day. And finally the Chickerings, cultured, amiable, and successful, sitting in their counting-house, the heads of a great and noble industry, the perfecters of a work of art that echoed in every tone the sweetness of Longfellow and the fancy of Hawthorne.

THE RISE OF THE NEW-YORK SCHOOL
OF PIANO-MAKING.

ALREADY the immense influx of German emi-
grants was making a radical change in American taste. The Boston school could not
satisfy a nation that worshiped the lusty genius of
Beethoven and Wagner. At the very time when
Chickering was perfecting his grand plate with its circular scale, a youth was growing to manhood in an
obscure German village who was to transform the
whole art.

Behind the career of Theodore Steinway we read
the charming story of his father, who, robbed of his
inheritance, orphaned and penniless at fifteen, made
his way by industry, integrity, and genius. Henry
Steinway's early youth was spent in military service.
Honorably discharged when twenty-one years old, he
began to learn the art of building church organs, and
to that end studied cabinet-making. His masterpiece
was no sooner complete than he married a woman
good, beautiful, and as poor as himself. In his leisure,
when a soldier, he had taught himself to make small
stringed instruments, such as zithers and mandolins;
but it was his heart's desire to become a piano-maker.
He supported his family by his trade, but, resolved
that his son should have the advantages he had not
been able to compass for himself, he devoted his

nights to making a piano upon which the lad should learn to play. It involved a year's labor, but its large, pure tone soon brought a purchaser, and made its maker's golden opportunity.

It is well known that the intelligent study of Bach begets a longing for great delicacy and, at the same time, fullness of tone. The double demand made upon his ear by the composer's two favorite instruments, the organ and the clavichord, has perpetuated itself in his music. In Bach you find the breadth and richness of the one, and the sensitive delicacy of the other. Henry Steinway, passionately fond of music, had in his early handicrafts exactly the same preparation of ear. You may trace in the piano which was the outcome of his genius, the artistic ideal of tone inherent in the music of Bach. The first Steinway piano ever made possessed it, and no temptation has ever carried its makers from the artistic aim it then discovered.

The years of Theodore Steinway's childhood were the preparation for the work of maturity. Henry Steinway, slowly raising himself from the straitened means of an artisan to the position of manufacturer, thought and talked at home of education, of music, and, most of all, of the art he loved. All these things became matters of reverence and ambition to his sons. Their play-room must literally have been the workshop, from the very beginning of memory; the difficulties and objects of emulation in their craft stimulated their curiosity. Its manual dexterities became their pride. There was the first-rate high school — the Jacobsohn Institute — where, till fourteen years of age, Theodore

Steinway studied; its excellent library he ransacked to feed his busy brain. There was a laboratory, too: he had a strong bent toward scientific pursuits; it was his delight to assist the professors in their experimental lectures. And there were the love and training of a mother to keep the home-nest warm and sweet.

It was a wholesome education. Every day the stimulus of a school that he was taught to regard as a privilege; the magic of science for his pleasure; the healthful craft of a noble art to train hand and observation; the travels and stories of the library to stir his imagination, and no money at all to buy the world and sell himself at once; but instead, unceasing economy, thrift, and industry in congenial occupations. As he went from piano to laboratory, from laboratory to factory, from factory to tales of geography and travel, his mind combined and compared. He resolved to see the world. He settled it then and there that the development of the science of acoustics would be the key to piano-making — that a perfect piano would be a perfect demonstration of the united laws of art and acoustics; and he rose every day with fresh determination to wring from nature the tone that hovered in fancy above romance and sonata, and had no existence behind the keys he vainly caressed. He dreamed of such wonders of sound as the world had never heard, and all these dreams he finally brought to pass.

What can be better for a lad than to love truly, work happily, and dream gloriously? There were other dreams and dreamers in that home at Seesen — mother-dreams of children reared with love and confidence in each other — and a little brother whose will, principle, talent for finance and business, was one day to build up the great manufactory that furnished Theodore Steinway the means and opportunity for the discoveries that have made him famous.

He had begun to study music at eight. In a house where music and the piano upon which to interpret it were the business of life, we may be sure that appreciation encouraged and discipline urged.

At fourteen he was able to play his father's piano in public, and his services had become so valuable that he could no longer be spared to school from the workshop. The occasion of this entrance into active work was a certain state fair in Brunswick, where Henry Steinway showed three pianos, and Theodore Steinway was sent to exhibit them publicly. The maker received the premium, and the boy praise,—sweet praise, for there was a real composer to judge and to give. Of all the pictures that grace the history of the piano, not one attracts me more than this: The father, soldierly in his bearing, among the crowd; the sturdy boy, his dress fresh from his mother's careful hands, bold through the grave responsibility of the hour, and fearless before his unaccustomed audience, because he played not to show himself, but for the sake of music and the family piano. He had helped to make the instruments he played; he had spent many an hour planning how by this or that stroke their beauties were most clearly brought out, their weak places (he would not have acknowledged them for worlds) concealed. He had felt the self-denial in-

15

volved in procuring the necessary capital; he had
heard and taken part in the discussion of every question in their make; and the ideas of science fermenting
in his brain had assuredly been broached at home and
put away for lack of money;—still, if not the piano of
his ideal, it was his father's, and he was proud enough
to do and suffer anything for it—yes, even to spend
his whole life in perfecting it, loving nothing else as
well to the day of his death.

So we can see him, with his life before him, playing in the little town hall at Brunswick. Then we
can follow the father and son walking home together
through the twilight. The door is open, waiting their
return; mother and sister meet the joyful pair and
listen to the story of the day's triumphs.

This was Theodore Steinway's valedictory to boy-
hood. Thenceforth his education was in his own
hands. What he made of it we know. But he lived
in a favorable age. The rapid and marvelous discoveries of science coincident with his lifetime intoxi-
cated the imagination. In the next years of patient
doing there was plenty of high and intellectual thinking; convictions strengthened and skill increased in
equal ratio.

"The sonority [of your instrument] is splendid and
essentially noble; moreover, you have discovered the
secret to lessen to an imperceptible point that un-
pleasant harmonic of the minor seventh, which heretofore made itself heard on the eighth or ninth node of
the longer strings to such a degree as to render some
of the finest chords cacophonic." So wrote Hector

Berlioz, when the International Exposition at Paris granted Theodore Steinway its gold medal. This reward of half a lifetime of devoted study came in 1867, twenty-eight years after the modest fair at Brunswick. It opened the way to the realization of his lifelong dreams. He wasted not one day before undertaking the experiments preliminary to a new system of stringing and building. His courage and his will rose for the struggle that would consume his remaining years of life, and sprang forward to victory.

Without some slight familiarity with the technique of piano-making, it is almost impossible to understand just what and how great were the difficulties which Theodore Steinway confronted. The slightest change in any comparatively unimportant feature of the instrument often alters the resulting tone, so as to involve a corresponding modification of every single element of construction. When he set about inventing the machine upon which to test the tension of his strings, he undertook at the same time to invent a totally new instrument. To augment the tension would involve discovering the corresponding alterations necessary in the soundboard which reproduced the vibrations; and also the necessary increase of size and weight (involving new problems of construction) in the hammers which struck these stiffer strings. If the hammer were heavier, the energy of the action must be increased without demanding more strength from the pianist. Furthermore, if the soundboard were made larger and thicker to bear the weight of the strings, it would be still less inclined to vibrate,

A Guarnerius Head and Neck.

and some means must be found to overcome its inertia. The huge iron frame demanded correspond-
ingly stronger wooden bracing, although, as the piano
was already too heavy, the total weight must be les-
sened rather than increased. To put it roughly, Mr.
Steinway was to double the energy of every factor by
which he excited or perpetuated vibration without any
addition to the initial force by which these energies
were set in action. But twenty-eight years of unremitting study and incessant practical experience had
been his preparation, and in the evolution of the piano
of 1867 many of the initial steps of future inventions
had been taken. The action-frame had already
been patented, and the ring-bridge promptly followed. These patents meant that he had made a
saving in the striking energy of his action, and had
augmented the means of setting his soundboard in
audible vibration.

Next came the experiments for the stringing.

"To string a violin correctly," says Mr. Hart, "is a
difficult undertaking. The first consideration should
be the constitution of the violin; strings that please one
instrument torture another. Neither Cremonese violins
nor old instruments generally require to be heavily
strung. The mellowness of the wood and their delicate construction require the stringing to be such as
will assist in bringing out that richness of tone which
belongs to first-rate instruments; if the bridge and soundboard be heavily weighted with thick strings vibration
will surely be checked." . . . "There are several
kinds of covered strings—those of silver wire have a

soft quality of tone; there are those of copper plated with silver, and also of copper alone, which have a powerful sound. There are violins that will take none but fourths of copper; there are others that will be simply crippled by their adoption."

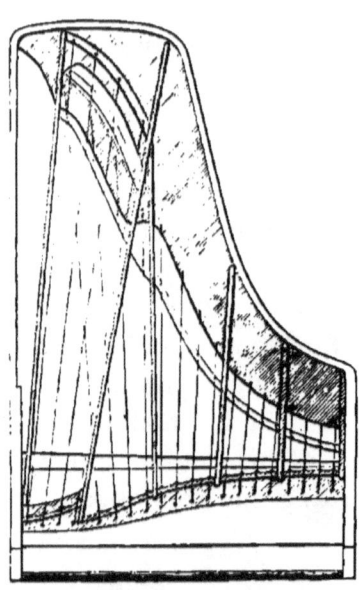

Original Overstrung Scale and Frame. Steinway & Sons' Patent, 1859.

Prior to the advent of the cupola metal frame all frames were flat, level, and rested upon extra strips of wood interposed between the soundboard edges and the iron frame. The cupola frame has its edges arched downward to the bent rim which supports the sounding-board.

So much for the fiddle and its four strings — but what of the piano with its eighty-eight keys, some of them controlling three unisons? The stringing and tension of a piano are problems ever present in the mind of its maker. That finally adopted by Theodore Steinway is the result of experiments and investigations with which he was incessantly occupied for at least three years. He brought to his task that gift of experimental dexterity which he had cultivated from childhood. Practically familiar with the demonstrations of vibrating strings made by the scientific world from Pythagoras to Helmholtz, in his factory at New-York he repeated and amplified them from the standpoint of the artist.

Three thick volumes, closely written, now in the possession of his family, record the completeness and number of his string tests and experiments. He created .his stringing, and ultimately his piano, in consequence of the discoveries then made, and that stringing is of such perfection that its vibrations have been photographed, a feat heretofore unparalleled in the history of piano-making.

Theodore Steinway found that a string stretched to its utmost limit of tenacity yields its largest, purest, and most brilliant tone. He calculated the tension involved, and resolved to bring every other part of his piano into symmetry. The single matter of the iron frame which was to resist this enormous strain showed the fiber of the man. The stoutest practicable frame of cast metal barely sufficed to stand a constant tension of twenty-five thousand pounds. He wished to carry its tensile strength beyond sixty thousand.

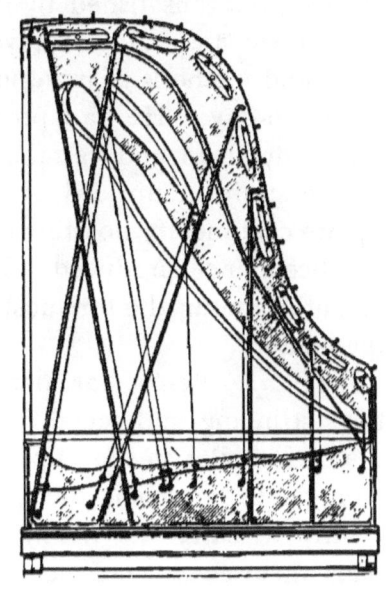

Condition of the Iron Frame in 1869.

No metal of the qualities which he desired had ever been invented. Enormous tensile strength was required, but this involved no corresponding necessity

for similar resistance to a crushing force. The demand
was unique.

In middle age he undertook a new field of study,
and began a pilgrimage among the iron-works of
Europe, as in youth he had sought its pianos.   Every
authority of the day was consulted, and every item of
practical experience eagerly collected.   He started
on his quest in 1869; a year and a half later he was
back in New-York, ready to experiment and invent.
The first patent for his new cupola frame was taken out
in 1872, but, unsatisfied by its success, he persisted in
experiment and study four years longer.   In 1876 his
alloy and methods of manipulation were completed as
in use to-day.   He had produced a casting that was
singularly clean and certain, and able to withstand a
tensile strain of upward of five thousand pounds per
square centimeter; no other cast-iron of similar strength
has been since produced.   Its achievement places its
inventor among the foremost practical metallurgists of
the century.

The *capo a'astro* bar, patented in 1875, completed
both stringing and metal framing; and the patent for
the action-pilot in the same year marks the successful
struggle to augment the energy of the action sufficiently to carry the heavy hammers.   But the hammers themselves were more refractory.   Two patents,
issued as late as 1880, show how sensitive and delicate
the adjustment required in this most important element
of the mechanism.

The grand case with bent rims was complete as
early as 1878.   To compass it he had built a new

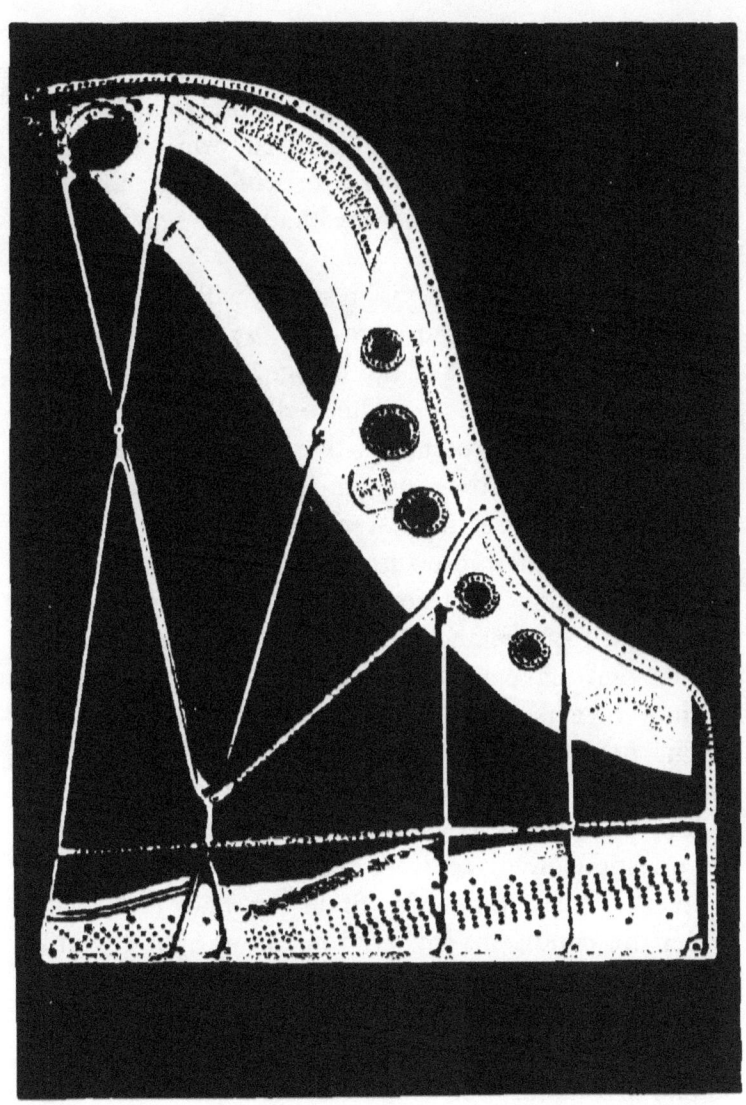

Top View of Theodore Steinway's Cupola Iron Frame.

sawmill to saw the veneers, sometimes twenty-three feet long, had discovered the proper curves in which to bend them, and had invented the machinery with which to press them into the required form. The mill and the tests for the braces, the iron frame and the two rims running different ways had cost one hundred thousand dollars.

The total tension of a violin strung with thick strings is $62\frac{3}{4}$ lbs.; its downward pressure, $27.13\frac{1}{4}$; the pressure on the treble foot of the bridge, 16.12 oz.; on the bass foot, 11 lbs. $1\frac{1}{4}$ oz. The total tension of a Steinway concert grand piano (with the scale given it by Mr. Steinway) is 60,000 lbs.; the pressure on the bridge is equivalent. Now, if most old violins certainly have suffered from their pressure of 27 lbs., and the belly has sunk behind the bridge, what is to become of the soundboard of a piano? And yet there are Steinway pianos that have satisfactorily borne their enormous strain for years.

The new tone pulsator evidenced the constant improvement in the soundboard, and the next season (1880) the composite soundboard-bridge showed that the questions involved were at last answered.

In the mean time Mr. Steinway had reproduced under the conditions involved in the upright construction each one of his discoveries. To him justly belongs the fame of making the American upright piano possible and successful.

The artistic instinct of the musician was always busy. The secret of his alloy in his grasp, he thought nothing of accruing fame; but his ear caught the altered

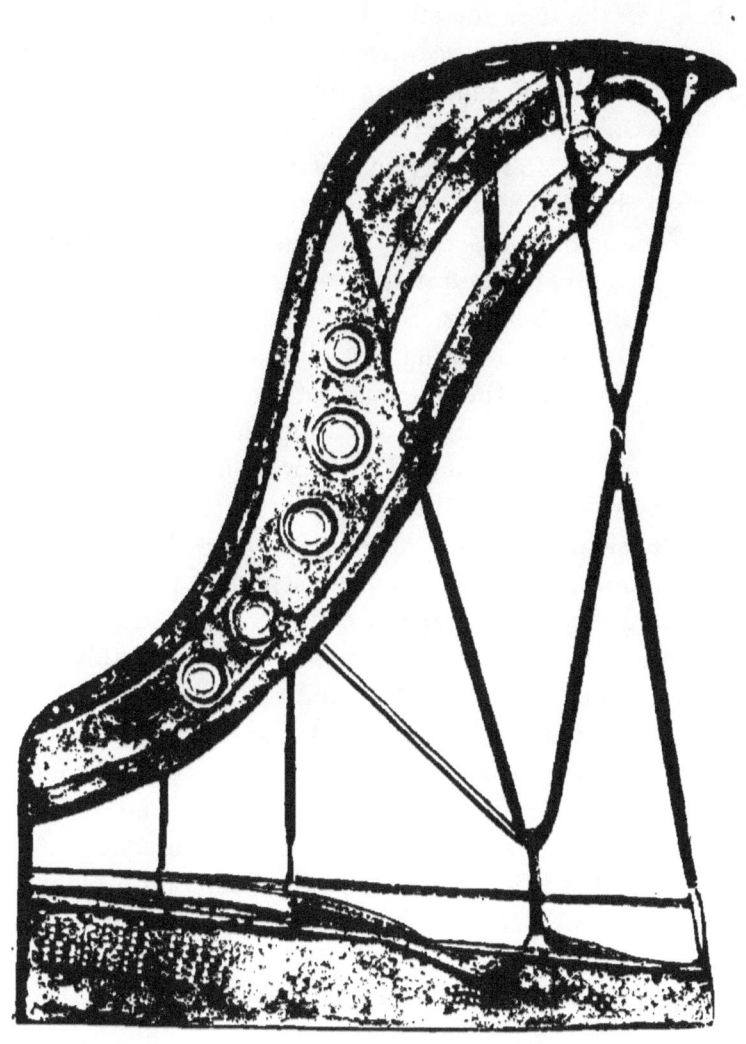

Reverse of Theodore Steinway's Cupola Iron Frame,
showing Inner Curved Surface.

clang of the new metal. In a former lecture I exhibited the peculiar tone-properties of vibrating metal;
I am also able to illustrate some modification of tone-
quality due to the presence of alloy. I sound this pair
of cymbals. You hear their dissonant clang. Listen
now to the richer tone of this second pair. The first
are American, the second Turkish. Western nations
have never yet penetrated the secret of this Oriental
brass. But when Mr. Steinway began to experiment
in steel alloys, in order to obtain a frame strong
enough to bear the pressure of his strings, he was
successful in producing an altogether new metallic
resonance. The clang of Steinway metal resembles
that of wood, *i. e.*, is pitched so low as to possess valuable musical vibration. Heretofore the chief anxiety
of piano-makers had been to separate the wooden
sounding parts of the piano from the iron, in order to
prevent the harsh metallic quality that invariably appeared when the metallic vibrations were audible.
The cupola frame itself, with its delicate upward
curve, was the result of the necessity of giving room
for vibration to the soundboard; and in setting the
iron frame it had long ago been discovered by Mr.
Steinway that it must not touch the board, from which
it was carefully isolated, but rest on the wooden frame
below it.

He now began to promote the elasticity of his iron
frame, and to seek a proper means of controlling it, and
at the same time conducting its vibration. Hence arose
his treble bell, which transmits the sound-wave to the
bent rim on which the soundboard rests, and so returns

The rim, braces, and key-bottom, in connection with the iron frame; the treble bell in place on the rim, at the left (seen from beneath).

the string-vibration to the board from beneath. The evolution of this artistic conception occupied twenty years. Valuable as it is in itself, its greatest use is in opening the way to future combinations and discoveries.

In 1883 Liszt wrote Theodore Steinway the last of a series of enthusiastic letters, telling him "your new grand piano is a glorious masterpiece, in power, sonority, singing quality, and perfect harmonic effects, affording delight even to my old piano-weary fingers." The great pianist and the great piano-maker, each nearer the end of his career than either guessed, rejoiced together over the new creation. Two years later the patent for the treble bell was taken out, and in this, the last important invention of the series, Theodore Steinway felt that the great work of his life was completed. An attack of pneumonia, too severe for even his immense strength, cut him down in the fullness of his powers and in the joy of his success. A widower and childless, his last thoughts were for the piano which his genius and devotion had created.

He was only sixty-five years old when he died, but his life had been rich and full. He had turned the local triumph of the Brunswick fair into the triumphs of world's international contests. The childish experiments in the Jacobsohn laboratory had led him, an honored member, into the world's great societies of fine arts, and the little piano with which his father's love had hoped to endow his childhood in an artisan's cottage, under the impulse of his genius, had become a king's treasure, and taken its place in the palaces of the mightiest monarchs of the globe.

THE VALUE OF PERSONALITY IN
CREATIVE ART.

THEODORE STEINWAY was a typical genius. With the financiering and management of the business which his father founded he had little to do.   Other brains and lives have built it up, and stamped it with their own personality. His work was of another sort, and his days were passed and his triumphs won in creating the piano — which his partners sold.   During the many months in which I prepared myself for these lectures by visits to the factories for practical study, the impression deepened daily that Theodore Steinway's piano was another body for his spirit, created by his personality.   The more intimately I acquainted myself with the secrets of its construction, the stronger grew my conviction that a man of equal genius but different personality could not have made it as it is to-day.   A narrower culture would not have conceived the broad artistic theory which united the world's experience in his treatment of the wooden portions.   A man unaccustomed to look with a traveler's eye over the whole earth would never have bethought him of the materials which he brought together to its composition. If unaccustomed to the precision of scientific demonstration, he could never have carried out the experiments which he successfully prosecuted.   Without mathematical and engineering skill he would have failed with action, iron frame, and bracing. It was the boldness of his imagination that conceived the outlines of his plans, and the unflinching persistence of his will that forced success in difficulties before which a weaker soul would have given up in despair. A less genial and sympathetic nature could not have obtained the necessary personal loyalty from the men destined to carry out his ideas, or the willing assistance of the specialists who gladly opened their secrets of science or skill to his eager gaze. His spirit infected the most opposite natures with an artistic reverence and zeal which still exist undiminished in the people who daily handle and live by the piano which bears his name.

Nor have the poetic aspect with which his imagination invested his art and craft, and the love which set a benediction upon the drudgery of its service, ever lost their hold upon the men whom he unconsciously taught to see what he saw, and feel what he felt. I never set foot among them, be they gentle or simple, without a direct consciousness of the being and personality of this man whom I never saw when in his mortal shape.

Every real piano leads captive the fancy of the mind which plans it, and the hold which a great piano has upon the affections of men long in its service is not to be uprooted by separation or change of worldly interests. The passionate devotion of employees of the Chickerings is proverbial, though Jonas Chickering, who first lit the flame, has been in his grave for forty years.

Theodore Steinway's personal habits dignified labor in every workman's eyes. When work pressed or went amiss, he was sure to throw off his coat and put things through with might and main; and keen must be the eye and sure the hand that could rival his. If the matter interested him, he chatted about it. In the wood-shop he spoke of wood, and in the action-room of levers.

His manly voice and his bright eye, his warm heart — a fullness of life that bounded in each pulse — left its impress on every man as he passed from one shop to another. And there was a character behind this. He had never tolerated the least shortcoming toward his art in himself; he had given all, and he asked neither less nor more of his men; and he received all he asked.

"I readily perceive with what enthusiastic love you seek to attain the incorporation of the most 'spirituelle' tone into the piano, which heretofore has only served as the exponent of actual musical sound. Our great tone masters, when writing the grandest of their creations for the pianoforte, seem to have had a presentiment of the ideal grand piano, as now attained by yourselves," writes Wagner.

Theodore Steinway's enthusiasm was so fervent, his impression of personal power so convincing, that he awakened a kindred spark in every workman, who felt himself honored by a moment of talk over his work. To-day, the highest mark of honor these men know how to pay me, as I go over that great manufactory, is to quote the chance comment of "Mr. Theodore."

17

When his nephews were old enough, into the shop every one of them went, and he must have had a talent for teaching, for they came out with equal reverence for the art and for him; and the one that lived the nearest to him was perhaps the one that loved him best.

The restless nature of the artist, held in check by duty and principle, came out as soon as means allowed. He seized the first opportunity to go over Europe, not as a tourist, but as a student. His services were in frequent demand as judge in the constantly recurring fairs and expositions in Europe. It was his duty, as his pleasure, to master every detail of the instruments entered for exhibition; and once seen, a piano was never forgotten. That narrow spirit that refuses to know or acknowledge a comrade's advance in his art was far from him. His own bold and original methods were the result of perfect familiarity with, and frank recognition of, every right effort. Hence he never wasted time in doing over what had already been well done, but kept in advance of his generation in both invention and research. " He knew every detail of every piano that has ever been made," said one of his nephews to me. He was all the time reading and thinking. Everything he heard or saw found its place and use in his art. As long as his pupils live the art will go forward in the loyalty and enthusiasm he infused into it. Determined and impulsive as he was,— and a more resolute will never worked in man,—he was quick to acknowledge and abandon a false position. The energy with which he prosecuted his in-

tentions made far less impression on his companions
than his perfect honesty with himself—the honesty of
the scientist whose most dreaded enemy is an illusion.
Full of his art, and never without some plan or experiment in it, he never prattled about his preoccupations.
During the years of experiment that preceded his
greatest discoveries, he said so little to those not actually working with him that the historical details of the
steps by which he advanced to success in many cases
died with him. His theory and his discoveries remain
a heritage to the art; but of the years of patience, of
self-discipline, of failure, and of anxiety that bought
his success, beyond the fact that they were lived side
by side with family and partners, there is little record.
The loneliness of genius was his also. Great as he
was to those who knew best the value of his work, his
own estimate of himself was humble. He had no time
to think about himself: the piano claimed all his imagination and his loyalty.

He was an omnivorous reader. The library that he
collected contains every sort of book, from a scientific
treatise to a French novel. The broadness of his
views sprang from sympathy with all sorts and conditions of human industry. His nature was rich and
genial, a tropical soil where principle had fostered the
noblest growth. He was capable of that rare thing—
tenderness. The men who, when children, provoked
his wrath by their boyish sins, dwell on the sweetness
of his forgiveness more than on the flash of his dis-
pleasure. His busy, skilful hand, his love of culture,
his patient industry, his royal generosity, won their

youthful worship, as his tremendous personal power, sincerity, and determination commanded the respect of the men he met in business.

His sense of tone was acutely developed; his critical ability in regard to touch and tone-quality was truly marvelous. Each word of his on this topic to those granted the privilege of listening to his discussion of this or that pianist's tone production, opened a new world of thought to the mental eye. His listeners seemed to become conscious of the very size and shape of the respective tones, so well was the theory of his subject mastered by this master of tone. He was able to produce the touch practically, too. His own touch was delicious.

The evolution of the piano, in his hands, was continuous through forty years: the germs of his greatest inventions, like the ring-bridge and the bent rim, appearing early in his career, and being traceable in every step till they came to perfection, not long before his death.

THE WORKING VALUE OF IDEALISM.

THEODORE STEINWAY brought the piano into harmony with mechanics, with science, and with art. He had equally pronounced talents in each direction, and, besides all, he was a poet. I do not speak of literary expression. Hamerton says that painters when asked for art criticism invariably send verses; and the musical and artistic instinct in Theodore Steinway sometimes beguiled him into rhyme. Beneath a habit of statement so uncompromising as frequently to be curt, and sometimes dogmatic, and a manner that never took on the gentleness and grace which early prosperity lends, there ran a strong vein of sentiment, and a rich, vigorous fancy. But these found their direct and natural expression in his art. They made him, perforce inventor and piano-maker. By poet I mean a man who is able to see the right relations of things visible to an ideal conception; or, as Emerson says, " the man who has the secret of the world where Being passes into appearance, and Unity into variety."

Emerson goes further, and tells us more: " Since everything in nature answers to a moral power, if any phenomenon remains brute and dark, it is because the corresponding faculty in the observer is not yet active. The path of things is silent. Will they suffer a speaker

to go with them? A spy they will not suffer. A lover, a
poet, is the transcending of their own nature; him they
will suffer. The condition on the poet's part is his re-
signing himself to the divine aura that breathes through
forms, and accompanying that. It is a secret which
every intellectual man quickly learns, that beyond the
energy of his possessed and conscious intellect, he is
capable of a new energy (as of an intellect doubled
upon itself) by abandonment to the nature of things;
that beside his privacy of power as an individual man,
there is a great public power on which he can draw by
unlocking at all risks his human doors, and suffering
the ethereal tides to roll and circulate through him.
Then he is caught up into the life of the universe.
His speech is thunder; his thought law."

There could hardly be a wider apparent difference
than that between Emerson the Seer, gentle, dignified,
fastidious, lean and ministerial, and Theodore Steinway, jovial, rugged, vigorous, passionate, with a will
and a personality that brooked no opposition from men
or things. But each had caught the same secret of
power. Each was a lover; each drew at will on the
great public power of the universe, for each had sympathy with the feeling, and invention, and personality
of all mankind. Theodore Steinway saw because he
was a poet; but he wrought because he was an artist—that is, a lover. Now, love is an energy whose
property it is to discover beauty, and to create it.
This is why love is said to be the fulfilment of law;
just as beauty is the revelation of law.

To insure a clear idea of a real lover, I quote the

definitions of a certain sweet saint, Thomas à Kempis, who made a life study of the subject :

> Love is the only thing that makes all burdens light :
> Bearing evenly what is uneven,
> Carrying a weight and not feeling it.
> Love knows no limit.
> It boils above all measure.
> It feels no weight, makes light of toil,
> Would do more than it can,
> Pleads no impossibility,—
> Because it thinks it can and may do all.
> So strong is it for everything, it everywhere
> Gives man a will to do work
> Where he that loves not faints and fails.
> In its vigils it may sleep, but yet it dozes not.
> Wearied, it is not worn.
> Bound, it is not confined —
> But, like a living flame, a burning torch,
> It bursts on high, and safely goes through all.
> If any man loves
> He knows what these words mean.
> A lover ought not to turn from his loved on account of
> contradictions.

Such is an artist's love of his art. Less falls below art into handicraft.

There is a passage in Ruskin that shows another side in this work: "The influence of the imagination over the senses is peculiarly sensible in the perpetual disposition of mankind to suppose that they see what they know, and vice versa in their not seeing what they do not know. 'The imitations of early art [he quotes Barry] are like those of children; nothing is seen in the spectacle before us, unless it be previously known

and sought for, and numberless observable differences
between the age of ignorance and that of knowledge
show how much the contraction or extension of our
sphere of vision depends upon other considerations
than the mere returns of our natural optics.'" Here, I
take it, is the common ground upon which all discovery in art or science rests; that inner image, clear and
strong, whose members and phases the artist presently
recognizes in the outward material world; for the
painter must have his technique, and pigments, and
perspectives, and anatomies; and the man of science
his laws and atoms; and the piano-maker his vibrations, and wood, and metal, and the like.

Theodore Steinway's inner sight was clear, his path
to his creation direct. No piano-maker ever wasted
less time in experimental failures. His piano shows a
gradual and consistent unfolding of his ideas, but very
few sporadic and abandoned hobbies. Thence the
absolute confidence with which he inspired those who
worked with him. "He knew!" they still say; the
phrase never varies. The ideal of the artist is a mental image of organized forms and colors; the ideal of
the scientist is an expected order of events; but the
ideal of the piano-maker is an ideal of tone, and since
no one but his inward self has ever heard that tone
till he haply brings it into outward being, his art is a
creative art; none the less an art because, like the
fabled angels of flame and of fire, it of itself chants
but one hymn.

It seems, at a superficial view, that the creation of
a musical scale, however wonderful its tone, is only

partial art, or at least incomplete art. I grant that the instrument and the player make the perfect completeness; and the great piano-makers were also players. But without the player, as the perfect instrument stands finished and waiting, its scale, that any unskilled finger can evoke, is, I take it, the very song that the morning stars sung together what time all the sons of God shouted for joy!

Mr. Steinway loved to travel. He was in the prime of life when he felt himself rich enough and free enough to make his first explorations. He went to Asia and Africa, and brought back a great store of information about customs and climates; specimens of wood; and a set of the musical instruments peculiar to each country. Then began the fine collection which he left to Hamburg. But in his fertile mind the collection was only a means to an end. Every one of those instruments possessed its own characteristic tone-quality, and was an example of one mode or another of exciting musical vibrations. When he had found a new instrument, he instantly formed an ideal of its possible perfection, and possessed himself not merely of the finished specimen, but of the history of its development, and the right principles of its construction.

The considerations that govern the tone of a stringed instrument, complex in application, are very simple in themselves. There are the tone strength and quality that lie in the material of the string, of its resonatory apparatus, and of its exciting means. There are the strength and tone-quality dependent upon the rela-

18

tive size of each of these three factors. There are the
strength and tone-quality dependent upon the proportions, curves, angles, and thickness of each and every
part, and related to these the shape and size of the air-
chamber, if any. But the infinite variety of combination proposes an endless number of problems. To
their solution the collector brought his quick eye and
quicker ear. He could separate without effort the
resonance of the wood from the hum of the strings,
and almost intuitively estimate the relative value of
each and every factor in the total result. His mind,
trained to scientific observation and deduction, stole
the secret of every instrument he saw; and the outcome was the clear and settled theory upon which he
made the revolutionary changes that have rendered
his piano different from all its predecessors. Those
who have the wit to make the ages their schoolmaster
learn well.

One of Mr. Steinway's favorite subjects of experiment, especially in the later part of his life, was the
violin — the most sensitive because the most perfect
of all. Not that he made his piano on the pattern
of the violin. He knew their common acoustic basis,
and was able to apply the discoveries and investigations made in one to the perfection of the other.

Mr. Hart thus sums up the construction of the
violin: "The back and sides are made of hard wood,
the belly of soft. The bass bar of soft wood not only
strengthens the instrument where the pressure of the
bridge is greatest, but is so exquisitely sensitive that
the slightest alteration in its position will effect such

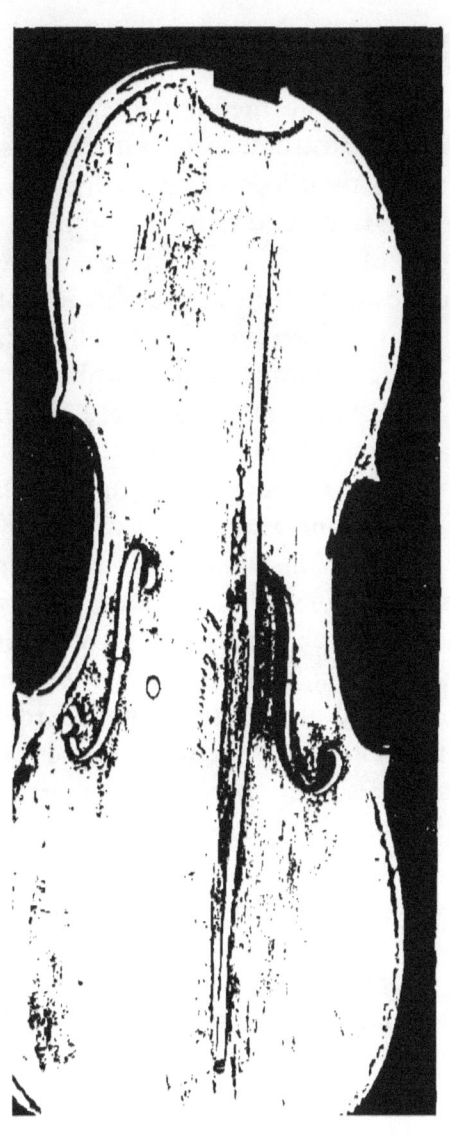

changes in the tone as often to make a good violin worthless.

"The sound-post is the medium by which the vibrating powers of the instrument are set in motion. It gives support to the right side of the belly. It transmits the vibrations, and regulates both the power and quality of tone. It is not possible to have a uniform position of the sound-post in all instruments. If the instrument is wanting in brilliancy, a bridge having solidity of fiber is necessary." . . . "The sound-holes are features of the greatest importance; upon the form given to them and the manner of cutting them largely depend the volume and quality of tone."

Every part of the violin finds its homologue in Theodore Steinway's piano: as, for instance, his bent rim, which answers exactly to the bent sides of the violin, and conducts the vibrations of the bridge all around the soundboard, as the sides of the violin carry them all around the back, and thereby set every part of it in vibration.

Sound-post.

The multiplicity of arts and sciences involved in the piano which Mr. Steinway made is very large. Many successful builders have begun as cabinet-makers, because, the piano being a wooden instrument, the knowledge of woods and their treatment involved in cabinet-making is a great step toward making a good piano. But something more than the mere study of wood as building material enters into the making of a musical instrument. Mr. Hart has much to say of the careful Italian violin-makers, who cut their wood into

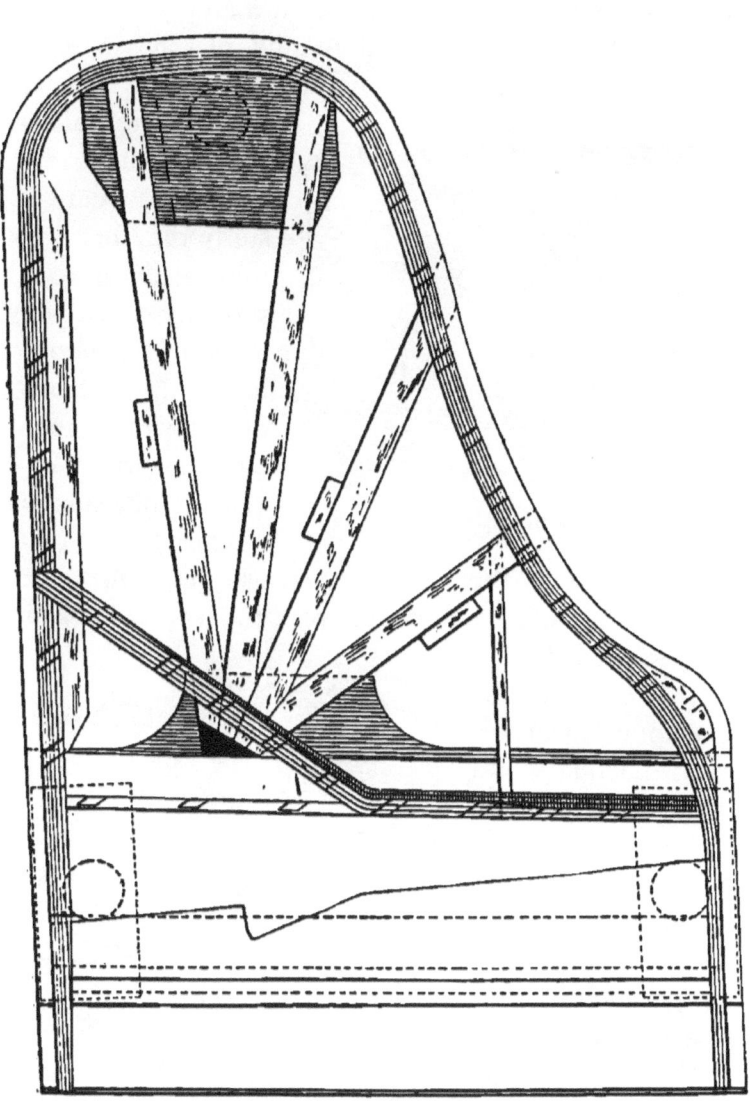

Theodore Steinway's Closing Rim. (From the
description of the patent.)

various-sized slabs so as to obtain a diatonic scale, and compared the intensity and quality of tone in each sample. "The scarcity of suitable wood was such as

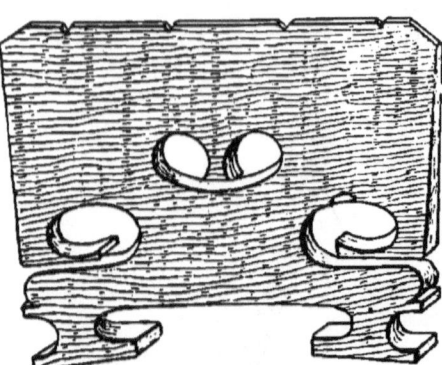

Vuillaume Model.

to make the makers hoard and use every particle. By careful study they brought the selection to a state of great perfection. The knowledge they gained of this vital branch of their art is enveloped in a similar obscurity to that which conceals their famous varnish; and in these branches of violin manufacture rests the secret of Italian success, and until it is rediscovered the Cremonese will remain unequaled in the manufacture of violins." . . . "It is my firm conviction that these great makers had certain guiding principles as regards the nature and

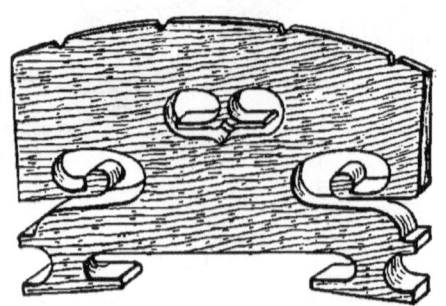

Bausch Model.

qualities of the wood they used, and that Stradivarius, in particular, made it a special study. The variation in the treatment of any particular instru-

The Back and Sides of a Violin.

Showing the four corner blocks, neck block, end block, and the linings.

H.C.BRONN.N.Y

ment depended on the difference in the quality of the material."

We have seen that Mr. Steinway traveled all over

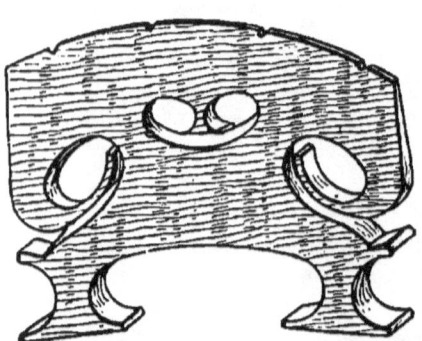

Old Italian Design.

the world, studying its climates, forests, and races. In his collection of musical instruments was embodied the artistic experience in the use and treatment of wood of each race and country he visited. The question of the curing and preparation of the wood used in his piano was one of the most constant and exhaustive of his investigations. It was owing to his success in this direction that the superb lumberyard at Astoria was begun, and the whole curing of the wood, from its first cutting to its perfection, taken into his own hands. These secrets of her own marvelous

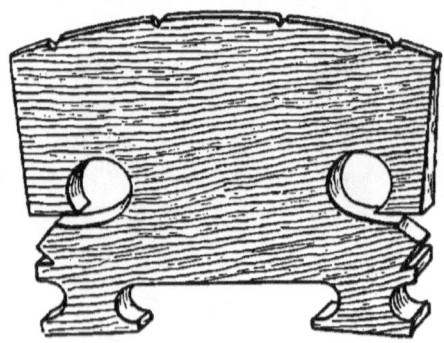

Common Bridge.

processes which he wrested from nature have, more than any other one thing, contributed to the charm of his artistic creation. Thanks to this knowledge, he has laid every zone under contribution; no mosaic was ever more delicately combined than the many woods that have each its best use in his piano. Among the twelve characteristic peculiarities of this piano, which he once furnished an inquiring friend, we read: "Through the careful choice of materials and careful construction of same, a duration of tone hitherto unknown was reached."

THE PIANO A REVELATION AND EXPO-SITION OF SCIENTIFIC LAWS.

BUT the wood, though the most delicate and perplexing part of piano-making, is only one item. Such a combination of materials, of metal, wood, and strings, as will produce the most powerful, brilliant, pure, and sympathetic tone is pre-eminently the feat of the artist and musician. While experiment, manual skill, and genius brought Stradivarius and Guarnerius del Gèsu to the perfection of violin-making, the piano is too massive and complicated a structure for artistic genius alone; it demands the exact calculations of science.

Theodore Steinway's work was the expression of the genius of his own generation. To understand it we must consider the train of thought characteristic of that generation, which has witnessed the evolution of modern science.

It is true that as early as the middle of the seven-

19

Stradivarius Model.
(Permission of G. Gemunder.)

teenth century Huyghens had proposed the undulatory theory of light, destined to play so great a part in the study of physical phenomena; and Hooke, by rubbing a card along the notches of a coin, brought to light the principle of the siren, that instrument which, in Helmholtz's hands, was finally to solve so many riddles.

Less than two hundred years ago Hawksbee demonstrated, by the silence of a bell rung in a vacuum, that sound is carried by matter. To-day it is easy for us to understand that various modes of motion, such as light, heat, electricity, and sound, advance in waves, as the motion of the ocean advances in waves. But the steps by which our knowledge came were very slow. We must recall that as late as 1870 Sir William Thompson thought it worth while to write a paper on the probability that the materials of the earth are really made up of elastic particles of matter, small but perfect, which scientists call molecules. He fixed their limits of size, from various calculations, at between one two hundred and fifty millionth, and one five hundred millionth of an inch; and, as an example, suggested that if we consider a sphere of water, as large as a pea, magnified to the size of the earth, the molecules that make it up would be coarser-grained than small shot, but not as coarse-grained as cricket-balls. We know to-day that a shock which sets these molecules quivering among themselves finally reaches our ears as sound. But, in 1853, all this, though not unthought of, was far from commonplace information. "Newton," writes Mr. Airy, "was the first to propose and work out the idea of a sound-wave"; and Tait credits Count Rumford with investigations which ultimately set going our modern idea of energy. It is not many years since sound was first defined as the result of motion—that is, a change in the relative position of particles.

The history of the steps by which men passed from the wave theory of motion, and the experiments of Rumford, to Sir William Thompson's conception of molecules, quoted above, is mainly the history of the investigations of motion, light, and heat—a history more wonderful than the Arabian Nights, but far too intricate to tell in this connection. I may only say that no molecule is supposed to be in actual contact with any other, and enumerate a few scientific achievements that most nearly touch the evolution of the piano.

Lichtenberg watched the motion of electrified powder strewn on electrified rosin-cake, and, in 1785, hardly more than a century ago, Chladni caught the idea,—investigated the motion of sand strewn on vibrating plates, and in so doing began the modern science of acoustics. It was toward the end of the same century that Newton calculated the velocity of sound in air; after him came Laplace, who corrected his figures. Dalton promulgated the atomic theory in 1808; it was taken up and worked out in the wave theory—the wave of molecular motion by which sound, like heat and electricity, advances.

In 1825, the year when Theodore Steinway was born, many physicists in Europe were measuring the velocity of sound in different substances. In 1827 Savart, best known by those famous experiments upon the scientific construction of the violin, discovered the three axes of elasticity in wood—along the fiber, across it, and along the annual rings.

In the first quarter of our century, Sir Charles Wheatstone, who had grown from a manufacturer of musical instruments into a savant, investigated the laws of sympathetic vibration. In these years, too, men like Faraday, Savart, and the brothers Weber were busy with Chladni's acoustical figures. Later Wertheim measured the velocity of sound through liquids, and Biot compared the velocity of sound in air with that in the iron water-pipes of Paris.

In 1842 Monsieur Marloye and other scientists were

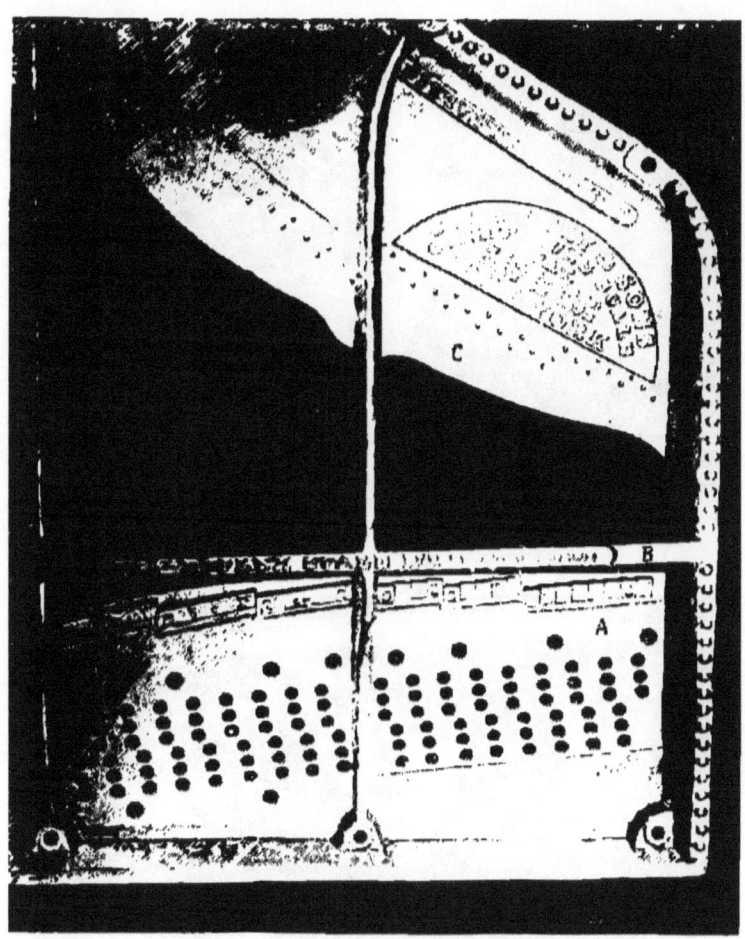

Treble Portion of the Iron Frame.

Showing A, wrest-plank plate, pierced for tuning-pins; B, *capo d'astro* bar; C, hitch-pin plate. The wrest-plank plate shows the front bridges, which support the strings and terminate their vibrating length.

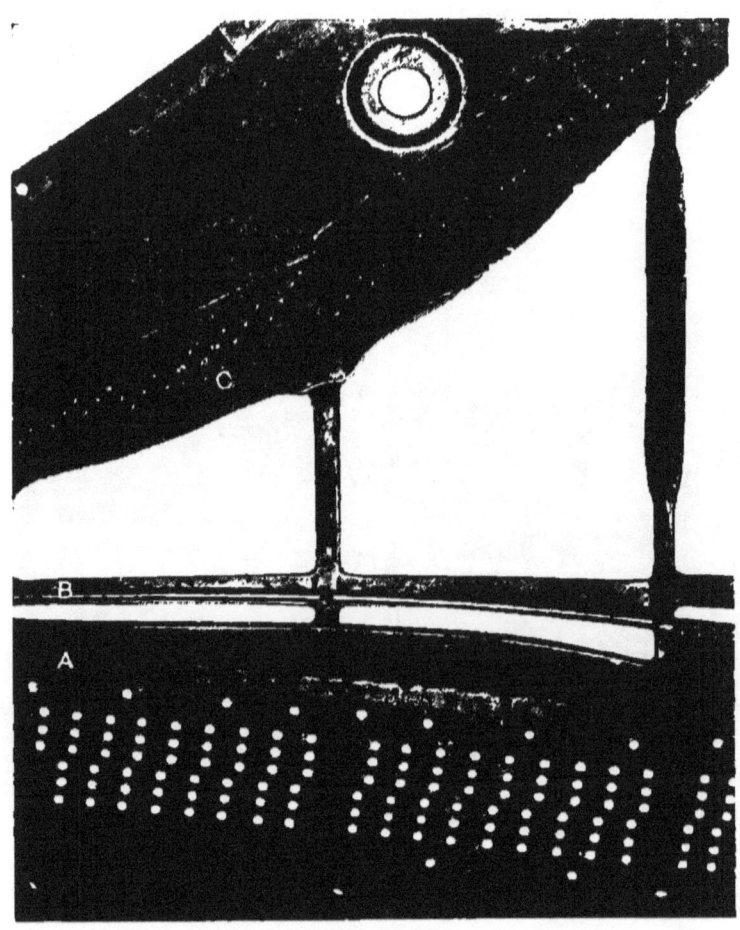

The Reverse of the Treble Side of the Iron Frame.

Showing A, portion of wrest-plank plate, drilled with holes to receive tuning-pins;
B, the polished edge of the *capo d'astro* bar; C, portion of hitch-pin plate. The
cut also shows two tension bars, and one rosette in the plate.

busy with the partial tones and partial motions of vibrating strings. It was in 1840 that Joule, in his investigations of heat, showed conclusively that the velocity of wave-motion increases with the elasticity of its medium. Wertheim carried forward Savart's work by demonstrating that sound, like other modes of motion, passes most strongly along the fiber of wood, less across the annual rings, and least along the annual rings—these directions coinciding with Savart's three axes of elasticity. But it was not till 1854 that Tyndall brought out his first paper on Molecular Influences; till 1859 that Helmholtz published his investigations on Nodal Points; or till 1862 that the latter brought out "The Sensations of Tone," which, with its recent revisions, has done so much to carry our knowledge of acoustics to its present point.

The scientific history of the piano, then, involves— (1) The sound-wave of Sir Isaac Newton. (2) The vibrations of plates, first studied by Chladni, and later by Faraday and others. (3) The atomic theory of Dalton, finally put into form by Sir William Thompson in his account of molecules. (4) The investigation of the laws of motion, molecular and otherwise— the work of the scientific world during the entire nineteenth century. (5) The velocities of sound in different media, determined in the first half of the century. (6) The different axes of elasticity in wood, investigated by Savart. (7) The longitudinal, transverse, and rotary vibrations of strings, studied by Marloye and others. (8) The laws of sympathetic vibration, investigated by Wheatstone. (9) The later and more complete exposition of the phenomena of sound-waves, by Helmholtz and his fellow-savants.

Each of these discoveries was studied by Theodore Steinway. Perhaps he discussed it with the very men who had made it. One by one he built them into his theory of musical sound and of the piano. Helmholtz, his friend and companion, the greatest and most ingenious of all acousticians, was his most stimulating influence. In many of the researches of "Die Lehre von Tonempfindungen" did the great piano-maker lend a hand; and Helmholtz's experiments on strings were performed on the piano that his friend created.

The debt of the piano to science has been disputed; and certainly Broadwood and his monochord mark the first deliberation which science and art ever held together over the matter. But Wheatstone, his curiosity awakened by the sympathetic strings of the viol tribe, presently combined both in himself; and Savart and Vuillaume, measuring and comparing and calculating, and experimenting with their Cremona fiddles, are stubborn facts. Theodore Steinway was a scientist from childhood; his methods of thought were trained in the processes of scientific reasoning; his mind was stored with scientific data. Every step in the conception and successive inventions of his wooden frame is founded on the work of Savart. His binding-bar, acoustic dowels, ring-bridge, rest on purely scientific considerations. None but a scientific mind would ever have seen the slightest analogy between the bent rim and the thin side of a violin. To a man who had no con-

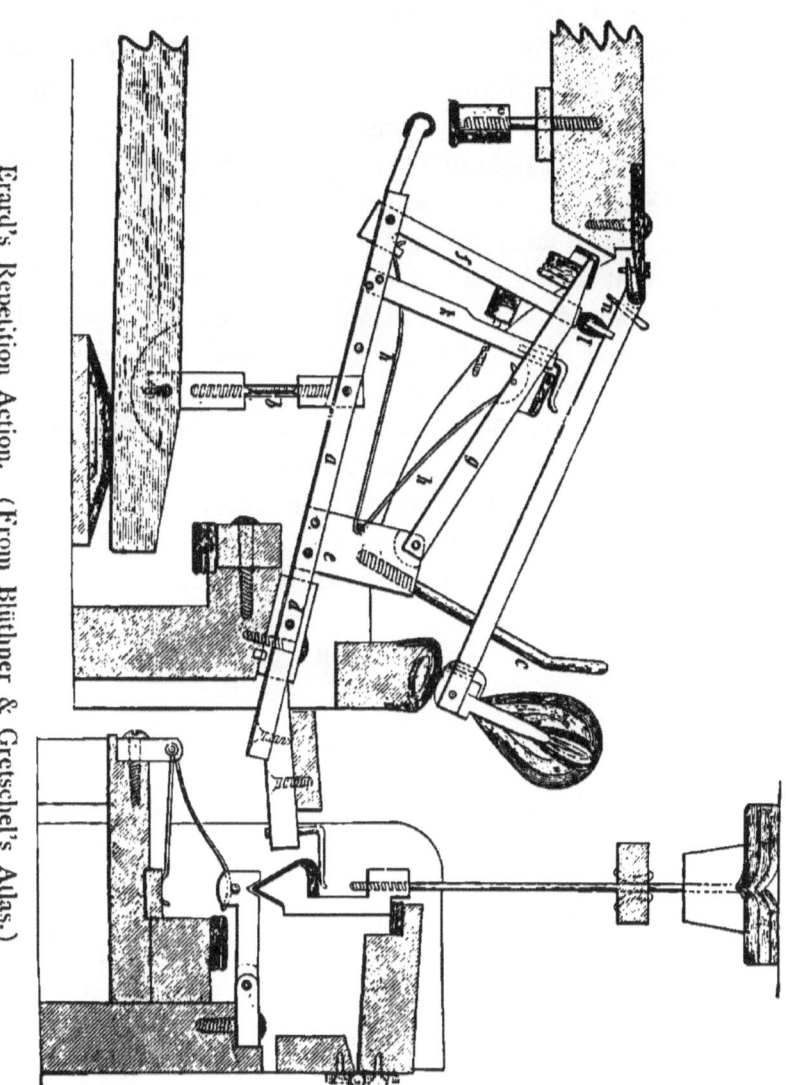

Erard's Repetition Action. (From Blüthner & Gretschel's Atlas.)

a, the jack-lever, technically, 'rairier'; *b*, the action-pilot; *c*, check; *d*, pivot of jack-lever; *e*, flange, which carries the jack-spring (*h*) and the pivot of the balance-lever (*g*); *f*, jack, controlled by spring (*h*); *k*, regulating flange of balance-lever (*g*); *i*, the knob (walze) on the hammer; *n*, escapement button, which controls balance-lever, and effects escapement of jack. The apparatus at the right belongs to the (under) damping. Observe the relative size of the extreme bass and treble hammers, drawn together on the plate.

ception of the nature of sound, and the progress of a
sound-wave, these things would have been impossible;
and what of his management of his soundboard,
which depends upon the difference between molecular and transverse vibration!

Similarly, the duplex scale is the fruit of Helmholtz's
researches into the scientific basis of harmony. The
ethereal quality of the una-corda pedal, depending on
sympathetic vibration,—in fact, the whole use of aliquot strings,—is an old inheritance from the viol. But
the *capo d'astro* bar in contact with the string at a
nodal point is a purely scientific conception. That it
was such to Theodore Steinway is clear, from an old
patent of his in which he mentions that the rotary
vibration of the strings on one side of the bar is in a
direction reverse to that on the other. The peculiar
tone-quality that he has evoked from his combinations
of wood and strings and hammers is the fruit of no
haphazard mechanic's genius.

> He had heard its faultless flow
> Where the roots of music grow.

So, too, the difference between the original Erard
grand action and the simplified and improved Steinway action in the piano arises from the application of
mathematics, to the solution of the questions involved.
Each of the changes by which Theodore Steinway advanced to sufficient lifting-power without making the
weight heavier, rests upon exactly such scientific considerations as enter into the calculations for astronomical
measurements, or architectural and engineering projects.

BELIEVE I have proved what I promised to demonstrate at the beginning of this lecture — namely, that *Piano-making is an art, because every true piano springs from an ideal of beauty, is a revelation of fulfilled natural laws, and as such embodies and expresses the personality of its creator.* In doing this it may seem that I have spoken too much and too warmly of the career of this one man. To develop the idea of the artistic quality in piano-making, it seemed necessary to show at least one artist at work, and the last great genius of the generation fast passing away, the man who has most formed and carried forward the art, was undoubtedly the one. But in a historical sketch like this I have found it impossible to take my text from any other piano.

For many weeks I haunted the patent-alcoves of the Astor Library, seeking the origin and tracing the development of the inventions with which we are familiar. Every great piano-manufactory in the world is represented upon its shelves in foreign and American patents, and between 1769 and 1826 the germ of almost every feature of our modern piano is recorded there, as the genius of successful invention and discovery made the round of Christendom. But the expedients which everybody uses, the settled traditions upon

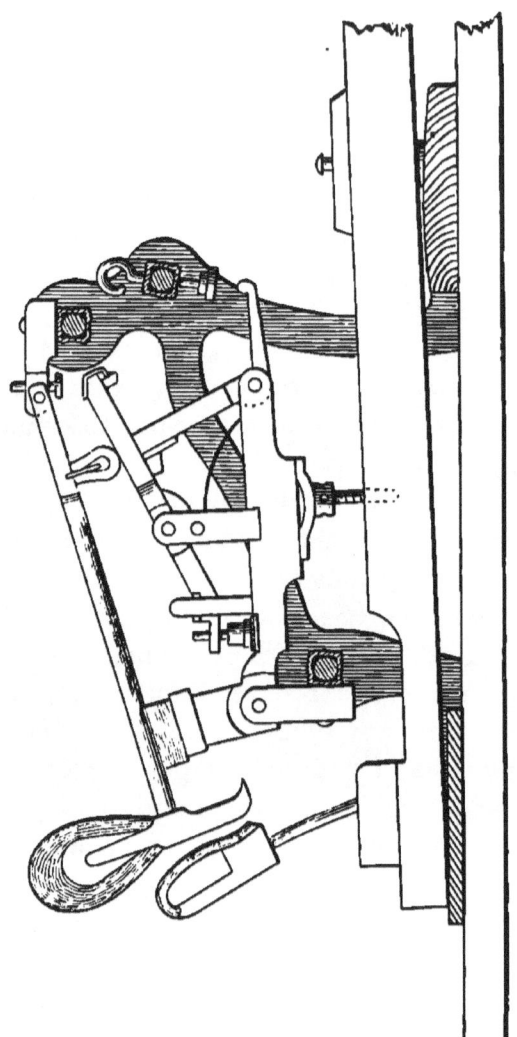

Theodore Steinway's Grand Action.

Showing free action-pilot, action-frame detached from key, and corrected and simplified leverage, etc.
The (over) damping is not shown.

which every German and American piano-maker works, mainly bear the image and superscription of Theodore Steinway. The iron frame has been worked into its present shape and bearing by his hand. The bridge on the soundboard, determined by the lengths of the strings, has the curve he gave it. Even as early as 1873 it was recorded in the official report of the Vienna International Exposition that "more than two thirds of the pianofortes there exhibited were imitations of the Steinway instruments"; and the jury passed a resolution of regret that the "path-breaking" firm of Steinway & Sons did not exhibit. The tone we hear —the characteristic American tone—has been enlarged and colored to suit the requirements of Theodore Steinway's ear. The web of diverging strings is his development of the Steinway fan-scale. I go to an action-factory, to see his action; I go to a hammer-maker's, to be shown his hammer. The finest workmen in America are proud to name the Steinway shops as their school of handicraft, and the majority of our best pianos trace their evolution directly to this factory.

We are now in the last decade of our century, which we like to call "America's age,"— the age in which we have made ourselves a great nation, and, as we proudly think, solved many a problem that the Old World hopelessly passed on to us. We survey our triumphs of government, art, science, and industrial invention; among them is certainly our own art of piano-making, in which we have outstripped all other nations, and formed a school for the world.

The struggle of the century wanes; its conquests
are made. Another generation, heir to the science and
skill of that fast passing away,— heir, too, to its genius
and its longing,— is already facing its own problems, and
marking out its own path. Its achievements will be
upon the foundation so well laid by the giants who
but yesterday folded their hands, their work complete.
Busy brains and fingers are carrying on the art whose
brilliant promise seems limitless, but the young men
and their conquests will be known as the glory of the
twentieth century.   Here at its threshold we halt,
draw breath, listen, ask — what of it all ? what of the
dead heroes of our art ?   Surely we too may hear

> Ascending pure the bell-like fame
> Of this or that down-trodden name,—
> Delicate spirits pushed away
> In the hot press of the noonday.
> And o'er the plain where the dead age
> Did its now silent warfare wage,—
> O'er that wide plain now wrapped in gloom,
> Where many a splendor finds a tomb,
> Many spent fames, and fallen knights,—
> The one or two immortal lights
> Rise slowly up into the sky,
> To shine there everlastingly.

THE END.

ACKNOWLEDGMENTS OF ASSISTANCE IN THE PREPARATION OF "A NOBLE ART."

So short a tract as "A Noble Art" permits no discussion of the principles set forth. But to the many gentlemen to whom the writer is indebted for the information upon which its statements are based she desires to express her thanks.

To Mr. Edward Porter, for much valuable knowledge of the special artistic principles involved in the practice of the Boston school of piano-making, of which he is an enthusiastic disciple.

To Mr. Morris Steinert, for access to, and information about, his fine historical collection of pianos and antique instruments; and also for the use of certain engravings in the present volume.

To Mrs. J. Crosby-Brown, for permission to copy plates in her superb work on "Musical Instruments"; and to the Crosby-Brown Collection of Musical Instruments in the Metropolitan Museum of New-York, which was of the greatest use in the preparation of the lecture on "The Evolution of the Piano."

To Professor Paine, of the Metropolitan Museum of New York, who kindly procured for the author an account of the Astor piano at Newburg.

To Mr. J. Howard Foote, who, on several occasions, placed his stock of metal musical instruments at the service of the author for her lectures, and procured information regarding the construction of the instruments.

To Mr. Henri Appy, violin virtuoso, and Mr. George Gemünder, artist violin-maker, for instruction in the principles of violin-making. Mr. Gemünder also furnished the author with the collection of violins, bridges, etc., etc., from which the plates illustrating the construction of the violin have been made.

To Mr. Bern. Boekelman, who furnished much information about the special beauties of European pianos, and on several occasions lent his clavichords and acoustic apparatus for the lectures.

To Mr. Samuel Johnston, the inventor of the "Johnston Harvester," who gave the writer much help in her researches in the manipulation and composition of iron and steel.

To Wessell, Nickel & Gross, who made a model of Cristofori's action, and would accept nothing for their trouble.

To Mr. Frederick Steinway, who lent the writer the photographic negatives which he had taken for his own use. The illustrations of the Steinway piano are prepared from them.

The following authorities have been consulted:

ACOUSTICS.

Helmholtz, Tyndall, Pepper, Faraday, Airy.

STRUCTURE OF WOOD.

Goodale.

HISTORICAL.

Carl Engel, "History of Musical Instruments"; Descriptive Catalogue of Exhibition at the South Kensington Museum, etc., etc.; Moscheles, "Recent Music and Musicians"; Fétis; Spire Blondel, "Histoire Anecdotique du Piano"; Naumann; Spillane, "History of the American Pianoforte."

TECHNICAL.

H. Schmitt, "Das Pedal des Claviers"; Blüthner & Gretschel, "Lehrbuch des Pianoforte-Baues"; S. Hansing, "Das Piano"; Rimbault; Brinsmead; A. J. Hipkins; Norton; Joseph Sittard, "Die Musik-Instrumente auf der Hamburgschen Aufstellung"; Grove's "Musical Dictionary"; descriptions of patents on pianofortes, in the U. S. Patent Office, to date; "The Metallurgy of Steel," by H. M. Howe.

VIOLIN.

G. Hart; Duborg; Fleming; Gemünder.